RAISING ANIMALS BY THE MOON

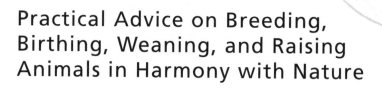

Practical Advice on Breeding, Birthing, Weaning, and Raising Animals in Harmony with Nature

LOUISE RIOTTE

STOREY BOOKS

Schoolhouse Road
Pownal, Vermont 05261

The mission of Storey Communications is to serve our customers
by publishing practical information that encourages
personal independence in harmony with the environment.

✿ ✿ ✿

Edited by Julia Needham and Doris Troy
Cover design by Carol Jessop, Black Trout Design
Text design by Mark Tomasi and Erin Lincourt
Production by Erin Lincourt
Production assistance by Elizabeth Collier and Jennifer Jepson Smith
Line drawings by Elayne Sears
Indexed by Susan Olason, Indexes & Knowledge Maps

✿ ✿ ✿

Printed in the United States by Versa Press
10 9 8 7 6 5 4 3 2 1

Library of Congress Cataloging-in-Publication Data

Riotte, Louise.
  Raising animals by the moon : practical advice on breeding, birthing, weaning, and raising animals in harmony with nature / Louise Riotte.
     p. cm.
  Includes index.
  ISBN: 1-58017-068-4
  1. Animal culture—Miscellanea. 2. Astrology. I. Title.

  BF1728.3 .R565 1999
  636 21—dc21
                                                                99-042376

C·O·N·T·E·N·T·S

*Astrology is not paganism or witchcraft,
nor is it a religion. But astrology and religion
are comfortable with each other. God made the
sun, moon, and stars; why should we not
be guided by His creations?*

— Louise Riotte
1909–1998

CHAPTER 1

COSMIC CLOCKS OF FUR, FIN, AND FEATHER

There shall be Signs in the Sun,

and in the Moon, and in the Stars.

— Luke 21:25

A strology is the science concerning the influence on earthly affairs of cosmic forces emanating from celestial bodies. And please let me stress my belief that knowledge of the celestial forces that govern the terrestrial sphere is true science, and the validity of astrology has never been disproved. Furthermore, science today is finding ample evidence to confirm many of the ancient beliefs and so-called superstitions. Astrology is *not* astromancy, which is fortune-telling. Do not *ever* confuse the two.

THE ZODIAC

All astrological considerations are based on the geocentric (having Earth as a center) position of the planets, on the theory that astrology is concerned with planetary motions only as they affect Earth. (See Resources for the annual astrological calendar from Llewellyn Publications, which is calculated, gives time changes for each section of the United States, and also tells how to alter the time for places outside the continental United States.)

The origins of both the calendar and the zodiac are almost inextricably woven together. Early humans had need of a time-measuring device and the conception of a zodiac was of immense help.

The word *zodiac* is commonly known but few among us can explain just exactly what it means. The word itself comes from the ancient Greek *zodiskos*, meaning "(a circle) of animals." Why, where, and when the animals were chosen is itself a fascinating story, but, alas, too lengthy to enter into here. What we are concerned with primarily is the practical, useful aspects of the zodiac and not its history.

The zodiac is a circle of 12 constellations, each of exactly 30 degrees, lying along the path of the sun. It is a tool by which the positions of the sun, moon, and planets can be measured.

Earth, rotating around the sun, is tipped at an angle of 23°27" to the horizontal, thus producing the alternating phenomena of summer and winter. Half of the zodiac lies to the north of the (celestial) equator and half to the south.

Nevertheless, as a circle has no beginning, early humans had to ordain a beginning from which celestial phenomena could be measured. Today this seems to be most readily provided by the equator, considered to be a great circle in the sky lying directly over Earth's equator.

It has been customary for centuries in the West to begin the zodiac from the vernal (spring) equinox. This equinox is the point at which the sun appears to cross the equator from south to north. This event occurs annually on March 20, 21, or 22. However, the equinox does not occur in exactly the same spot two years in succession. Instead, its place slowly rotates around the sky, taking about 25,800 years to complete the circuit. Astrologers call this phenomenon the Precession of the Equinoxes.

In consequence, the beginning of the (tropical) zodiac, generally known as the First Point of Aries, also moves slowly backward around the sky at the same pace, amounting to 50 seconds of arc annually. Normally measured from the First Point of Aries, the zodiac is a moving one, just as the equinoxes and solstices are moving points, recording astronomical events that never take place in exactly the same point on the sun's path for two years in a row.

THE SIGNS

There are 12 signs in astrology, and, as it circulates through the sky, the moon passes through each of them in succession. These 12 signs are:

Zodiac

- 🐂 *Aries, March 21 to April 19,* ruled by Mars: A movable, fiery, barren, masculine sign. Governs head and face.
- 🐂 *Taurus, April 20 to May 20,* ruled by Venus: A fixed, semi-fruitful, earthly, feminine sign. Governs throat and neck.
- 🐐 *Gemini, May 21 to June 20,* ruled by Mercury: A flexed, airy, barren, masculine sign. Governs hands, arms, shoulders, lungs, and nervous system.
- 🦀 *Cancer, June 21 to July 22,* ruled by the Moon: A movable, watery, fruitful, feminine sign. Governs breasts and stomach.
- 🐂 *Leo, July 23 to August 22,* ruled by the Sun: A fixed, fiery, barren, masculine sign. Governs heart, sides, and upper portion of the back.
- 🐐 *Virgo, August 23 to September 22,* ruled by Mercury: An earthly, flexed, barren, feminine sign. Governs solar plexus and bowels.
- ♎ *Libra, September 23 to October 22,* ruled by Venus: A movable, airy, semi-fruitful, masculine sign. Governs kidneys, loins, ovaries, and lower portion of the back.
- 🦂 *Scorpio, October 23 to November 21,* ruled by Mars: A fixed, fruitful, watery, feminine sign. Governs sex organs and the bladder.
- 🦌 *Sagittarius, November 22 to December 21,* ruled by Jupiter: A fiery, masculine sign. Governs liver, hips, thighs, and condition of the blood.
- ♑ *Capricorn, December 22 to January 19,* ruled by Saturn: A movable, earthly, feminine sign. Governs knees.
- ♒ *Aquarius, January 20 to February 18,* ruled by Uranus: A fixed, masculine, airy sign. Governs calves, ankles, distribution of bodily fluids, and intuition.
- 🐟 *Pisces, February 19 to March 20,* ruled by Neptune: A flexed, watery, fruitful, feminine sign. Governs feet and the psychic faculty.

Besides the foregoing indications given by the signs of the zodiac, each planet produces its own effects.

AFFINITIES OF PLANETS

The moon rules the fluidic (watery) elements of the body and also the brain matter; thus, all lunar aspects are particularly influential on the body of humans or animals through these sources, that is, through the brain and the chemical constituents of the body. This includes the stomach and the breasts.

Solar (sun) aspects are effective through the cerebellum (the portion of the brain that lies in the lower back part of the cranium), the heart, sides, and the upper portion of the back.

Other planets are particularly influential on the body through the organs with which they have vibratory affinity and, because of this, are said to rule them.

Mercury affects the sympathetic system, the nerves, bowels, hands, arms, shoulders, and lungs.

Venus is effective over the throat, lymph, skin, kidneys, and ovaries.

Mars influences the muscular system, bladder, sex organs, and the head.

Jupiter influences the condition and action of the liver, blood, and thighs.

Saturn affects the spleen, knees, and bony structure.

Uranus influences intuition, as well as the ankles and the distribution of gases and fluids in the body.

Neptune influences psychic emotions, bestowing mediumistic and psychometric gifts or tendencies, and also influences the feet.

Pluto is the ruler of the underworld, or the subconscious part of the body. Many astrologers consider Pluto the ruler of Scorpio, the eighth sign of the zodiac, which relates to death and rebirth.

THE INFLUENCE OF CELESTIAL BODIES

Our main interest lies in the influence of the celestial bodies on our lives and on those of our domestic animals. And the body of most interest is the moon, which directly influences water. We are all familiar with the effect of the moon on the tides. What we may not realize is the extent to which our own biological rhythms are affected by the moon. About seven-tenths of the adult human body is made up of water. It is the water in our bodies, and in the bodies of animals, that makes us, and them, responsive to the moon.

Every plant, animal, fish, or insect responds to exogenous (exterior) forces, for the most part unseen and unknown. We are also influenced by many biological cycles. Blood pressure, temperature, and even moods rise and fall at regular intervals, and these cycles are all variable and begin at birth. Animals are influenced in the same way; we call the effects instinct.

All animal life is affected by the weather, and it has been repeatedly proved that sunspots and solar flares have a powerful effect on Earth's weather, particularly affecting the ion concentration in the atmosphere.

Human, animal, and plant life are all affected by electrical atmospheric changes, geomagnetism, and electromagnetism. It has been proved that animal life is affected by the positive and negative ions in the atmosphere. Today we also know that cosmic conditions slow down, speed up, or even halt many biochemical processes.

Over and over again it has been demonstrated that the ever changing angles formed by the planets trigger solar flares and spots. There is even certain evidence that they are also responsible for seismic and volcanic disturbances, due to the shifting of the center of gravity of the solar system.

Naturally, since planetary positions affect solar magnetic field phenomena and the magnetic field patterns of Earth, they may be expected to cause some reactions among the creatures that inhabit our planet. We must conclude, therefore, as did both the ancients and modern science, that the harmonic angles formed by the planets actually do affect all life forms.

This knowledge, put to practical use in caring for livestock, pets, bees, and fish, can give us rich rewards. We may expect our animals to have better health, grow and mature more rapidly, be more productive (milk animals), and improve as breeding stock.

USING THE SIGNS

The signs have many practical applications. For instance, weaning young farm animals will not be a problem if the proper sign of the zodiac is observed. When the sign is disregarded, a mare will whinny and fret and may even sicken. Her foal, too, will not do well; at best, it will be slow in learning to eat. But if the foal is removed from the mare when the moon is in the signs of the zodiac that do not rule the vital organs of the body (Capricorn, Aquarius, Pisces, and Sagittarius), the mare will scarcely look up from grazing and the weanling will flourish, learning quickly to fend for itself.

When we learn the laws of nature — what to do and when to do it — we have the proper tools for strengthening both ourselves and our stock and curing or preventing many ills. Furthermore, working with nature's good influences gives us confidence and improves our ability to achieve what we set out to do. Well informed, we are no longer the playthings of every ill wind that blows, for we can intelligently use and manipulate the vibrations emanating from the planets, transforming or transmuting them into opportunities for the advancement of our goals.

BREEDING

If you intend to breed any livestock, the rules of astrology are especially important for you to know. Here are some general principles, and you will find more information in the chapters on care of particular animals.

Take care when setting eggs and mating animals that the young will be born in a fruitful sign and an increasing moon. The most fruitful signs are Cancer, Scorpio, and Pisces. The young born during these signs are healthier, will mature faster, and make more satisfactory breeding stock. Those born during the semi-fruitful signs of Taurus and Capricorn will also mature quickly, but generally produce leaner meat. Those born in the sign of Libra are more likely to be beautiful, graceful, and of good disposition. This is the important sign for showing and racing.

To determine the best date to mate animals or set eggs, subtract the number of days for incubation or gestation from the fruitful dates given above. For example, cats and dogs are mated 63 days prior to the desired birth date, as shown; chicken eggs are set 21 days prior.

The annual *Moon Sign Book and Gardening Almanac,* put out by Llewellyn Publications, gives the best dates for setting chicken eggs by months (in addition to the best dates to dock, dehorn, and castrate animals).

✩✩ GESTATION AND INCUBATION FIGURES ✩✩

ANIMAL	NUMBER OF YOUNG	GESTATION
Goat	1 or 2	151 days
Rabbit	4 to 8	30 days
Cat	4 to 6	63 days
Dog	6 to 8	63 days
Horse	1	346 days
Cow	1	283 days
Pig	10	110 days
Sheep	1 or 2	150 days
Guinea pig	2 to 6	62 days
Hamster	1 to 12	15 to 18 days
Domestic Fowl		
Chicken (hen)	5 to 20 eggs	21 days
Turkey	12 to 15	26 to 30 days
Guinea hen	15 to 18	25 or 26 days
Peahen	10	28 to 30 days
Duck	9 to 12	25 to 32 days
Goose	15 to 18	27 to 33 days
Pigeon	2	16 to 20 days
Canary	3 or 4	13 or 14 days
Budgerigar	4 to 8	18 to 21 days
Lovebird	4 to 8	18 to 21 days

ASTRONOMICAL AND ASTROLOGICAL TERMS

The following are some terms you will find used in this book. For more information, interested readers should consult some of the works listed in Suggested Readings.

✩ *Ascendant.* The degree of the zodiac that appeared on the eastern horizon at the moment for which a horoscope is to be cast.

✡ *Aspect.* An aspect is a set distance between planets, or between a planet and the ascendant, or between a planet and the midheaven. Aspect is applicable to any blending of rays that results in their interactivity. The body with the faster mean motion is said to aspect the slower.

✡ *Conjunction.* The moment of closest approach to each other of any two heavenly bodies.

✡ *Equinox.* One of the two times in the year when the sun crosses the equator, making day and night of equal length in all parts of Earth.

✡ *Opposition.* The time when the sun and the moon or a planet appear on opposite sides of the sky.

✡ *Rules.* Has dominion over; influences.

✡ *Solstices.* The points in the ecliptic at which the sun is at its greatest distance north or south of the equator. The *summer solstice* occurs when the sun is at 0° Cancer, about June 21; at the *winter solstice* the sun is at 0° Capricorn, about December 21.

✡ *Sunrise and sunset.* The visible rising and setting of the sun's upper limb across the unobstructed horizon of an observer whose eyes are 15 feet above ground level.

✡ *Trine.* An aspect of 120°.

✡ *Twilight.* This time begins and ends when stars of the sixth magnitude disappear and appear at the zenith, or when the sun is approximately 18° below the horizon.

✡ *Zenith.* Mathematically, the pole of the horizon.

WHAT SIGN IS YOUR PET?

This can be one of the more interesting aspects of pet ownership. If you buy a purebred animal, there will, of course, be records and you can check back to see when the animal was born. But what about that charming cat or kitten who suddenly appeared one day at your back door, or that lovable mutt you couldn't resist at the dog pound? Is there any way you can determine their signs? In general, yes.

Aries

Aries animals are very athletic, and their excessive vitality may cause them to eat large quantities of food. However, they rarely become fat and their muscles fairly ripple their hair with vibrant energy.

Most likely the animal will have short hair — even cats such as Persians and Angoras born in Aries will have shorter hair than others of their breed, with reddish brown coats (dogs may have this as a base color). Domestic animals such as cows and pigs will have a stronger and slightly chunkier build, reflecting their rugged constitution.

Notice the voice: Your cat's booming meow is likely to be in the key of C, and your dog will have a deeper growl than his peers, reflecting the Martian influence. Even your cow, bull, or pig will have a deeper "voice."

Taurus

Taurean animals have a "finished" look even when first born, and never lose it throughout their lives. You may expect your cat or dog born in Taurus to be better-looking than others of its breed. If it's a show animal, it will win points for a finely configured face. However, keep your Taurean on a strict diet; there is a strong tendency to overeat and become too heavy.

Medium-long hair, either a very pale brown or of a golden sheen, is characteristic of Taurean animals. Particularly noticeable in your cat is a high-pitched voice — and he may sometimes call you in a piercing high A! Taurean birds are good singers.

Gemini

Gemini cats are quite likely to "warble" rather than meow — that's both sides of the twin in her trying to come through at the same time! If she could sound the tonic chord, she would be on her way to fame and fortune; since she can't, she's probably selected E as the best compromise.

Your Gemini pet may have either a long coat or short hair, or it may even be somewhere in the middle. She doesn't care, as long as it isn't shaggy and doesn't impede her speedy progress, for this animal's body is quick (and her mind is even faster) and she'll tumble over herself in an effort to direct her body to where her mind chooses to send it next. If your pet is a goat, you must think fast too — very fast!

Geminis feel the same way about color — they don't really care. However, if offered a choice, your Gemini would probably settle for white or perhaps a slate gray. You may expect her to be reasonably well behaved around others of the same breed and she may even exhibit a genuine liking for them — a trait that is important in herd animals.

Cancer

Cancer animals are likely to have a coat with white or silver gray as its most prominent color — and the hair is apt to be longer even on the short-haired breeds, such as Burmese and Siamese in cats and the Chihuahua in dogs. Cats can be expected to have blue or green eyes, according to their breed. Voices are probably a pleasing contralto.

Animals do empathize with their humans — your dog will suffer when you suffer and perhaps somatize your illness when you are sick. He will always do his best to love and protect you — a strong trait even in a Cancerian horse. Cancerian animals are fiercely loyal. They are also good sentries and excellent protectors of children. If your animal exhibits these traits, it is quite likely a Cancer.

Leo

Leo animals have coats tending toward the solid golden hues, and this may be particularly noticeable in the coats of horses known as palominos, although those colors show up in any breed. Leo cats are almost never calicoes, and neither cats nor dogs are likely to have mottled fur. Their sleek coats are evidence of good health, though animals born under this sign may have slightly smaller bodies and "pointy" ears. The cat's midnight roar will most often be in the ferocious key of D, and your dog will also have a deeper bark that may at times surprise you. Bulls will bellow louder than most others of their breed, and your horses will "nicker" in a deeper key.

A tendency toward aggressiveness may be exhibited in Leo females, but they can put up with each other more readily and do not fight as often as males. Leo animals are known for their beautiful eyes.

Virgo

Virgo animals are not athletically inclined, and a racehorse born in Virgo seldom makes a good long-distance runner (Aries is the sign for stamina). To some extent, this is a matter of physical structure as well as disposition, for short, stubby legs are the rule. Animals under this sign, especially dogs, also find it difficult to range the field, as their long coats pick up nature's debris along the way: Dirt, leaves, burrs, and seeds seem to cling to this animal more than to the coats of other signs. On the other hand, the short legs and compact body may be an asset to meat-producing animals such as cattle and rabbits.

The dogs, oddly, have a dainty, chirping bark and the cats a 20-watt *purr-rr-rr-rrrr*! Your canary is more likely to be a "chirper" than a singer.

It was probably a Virgo cat who started the tradition of having "nine lives." An animal born under this sign gets into endless mischief and seems to survive almost any mishap, ready to go right on to the next "stunt"!

Libra

Libra animals probably began another tradition — that of being "born beautiful." If you are in doubt as to your pet's sign, try this as a yardstick. Libra animals are not only beautiful physically but also agreeable and charming in disposition. They also charm you: The dog constantly aims to please and even the cat will unbend a little and make a greater effort to be friendly. They also seem to have an extra measure of pride, often demonstrated by the extra measure of poise with which your Libran greets his fellows. You can instantly pick him out of a crowd, for his head always seems to be just a bit higher — and this reflects his healthy self-image. At home he will unerringly pick a background suitable to his beauty, one that "shows him off." A white Persian loves to lounge on a bright red or blue pillow. Your golden retriever may choose a green chair to complement his lovely color. Give your Libran a jeweled collar and just watch how proudly he wears it.

The natural element for Librans is air, so observe how he walks — or rather floats — through the living room. Yes, floats: Librans don't walk or saunter or move about like other cats. Rather, they appear to glide above the surface, much as a sleek catamaran slides over the water. Your Libran will be especially pleased with pile carpeting about 3 inches thick, with a thick foam pad beneath, for this will serve to intensify the illusion he chooses to create. Turn your pet bird loose (but not with the cat in the same room), and his graceful flight will amaze you.

Scorpio

Scorpio animals are very sexy, the females in particular having a strong attraction for the males. If your dog or cat isn't spayed, be sure to keep her penned when she is in season, or you may have more animals about than you bargained for!

Both dogs and cats (as well as other animals) are likely to have a reddish coat (in horses this color is called roan), with a golden sheen that may appear to be more an amalgam than one original and solid shade. They are often quite beautiful if well brushed.

Voices tend to strike up a middle C and go on from there.

Cats born under this sign need to drink lots of water and some milk. Occasionally give your dog beef broth — it's good for him. This is necessary because, zodiacally speaking, Scorpio animals may run afoul in their more "southernly quadrants."

Sagittarius

A Sagittarius animal is likely to have a large body on rather stubby legs. Its coat, if the animal is a cat, will be neither pure tortoiseshell nor calico, nor even tabby, though more nearly resembling the latter. Both dogs and cats (as well as other animals) are apt to have somewhat blunt-looking faces. Sagittarius animals (especially horses) are strong and healthy but have a tendency to hip and thigh problems, especially if allowed to range freely. Should an eye, kidney, or circulatory problem develop, be alert and take your animal to a vet — don't let it run on in the hope that it will get better.

Unless spoken to, your Sagittarian will rarely speak, but when he does, his voice is surprisingly soprano for all its comparative large size. Sagittariain animals are restless, especially cats. No matter which side of the door he is on, your pet will want to be on the other side — provide him with a "kitty door" and then he can go in and out as he pleases. Unlike the Libran, the Sagittarian animal is inclined to be somewhat clumsy, so don't leave fragile ornaments on small, tipsy tables. Fortunately, he has a hard head and is seldom hurt when something comes crashing down on him.

Capricorn

Capricorn-born animals are likely to outdo others of their breed in capriciousness and athletic activity, and most have long lovely legs well suited to their tendency to prance about. This active animal will appreciate lots of toys. Ideally Capricorn coats are short and sleek and most often brown, possibly a warm, chestnut brown that is pleasing to the eye. Watch out for knee troubles; those lovely knees are prone to injury — pay particular attention to this if your pet is a horse. Capricorns are graceful but as they travel about, brushing against one thing and another, they are subject to another Capricorn disorder — skin troubles. If this develops, get your pet to the vet for a gentle ointment. Capricorn-born goats are mischievous and very active indeed!

If your pet is a cat, expect her to vibrate in F!

Aquarius

Aquarians are generally unconventional in nature, and probably the most obvious point will be the way your pet wears his coat. If others of his breed have short hair, his will be long. It may also have a tendency to be wavy and even at times to be tightly curled — take particular note of this if your pet happens to be a sheep or a dog or another animal with this type of hair. Aquarius pigs have very curly tails!

Aquarians like to make their homes on boats. If you have a small craft laid up in your yard or garage over the nonboating months, don't be surprised to find your Aquarian sleeping in it. Come spring he will be reluctant to leave it — take him along when you launch it back into the water.

Your Aquarian cat will be frightened of your vacuum cleaner — to him it's a roaring monster. Other cats may take it casually, but your Aquarian will scoot under the nearest chair or bed — it's one Unidentified Terrestrial Object he doesn't want to tangle with.

Aquarians like to eat frequently and well. Expect your pet to have a hearty appetite, but Aquarians are so active that they seldom gain too much weight. Aquarian cats make good fathers and, more than those born under any other sign, are quite likely to baby-sit the kittens when mother goes to eat or to relieve herself.

Their excessive activity can lead to trouble; Aquarians have weak legs, calves, and ankles — watch for this disability and care for it as soon as you notice it. Otherwise, they have good health and are likely to have a long and happy life.

Pisces

Pisceans have a strong instinct for self-preservation, and this keeps them from some of the external difficulties that may plague animals of other signs. The major exception is the wear and tear their feet get as they dash madly about. Keep them at home if an asphalt street is being laid in your neighborhood — cleaning them up can be a tedious job.

Nature did a marvelous job of confusing your Piscean both physically and mentally. His coat, although short, will be "double" (one for each personality?) — thick underneath and less dense but very soft on the surface. I identified my own white cat, Jacob, because he wore this type of coat — he is definitely Piscean. A white coat, or one of muted plaid, is typical of Pisceans, and they are likely to have blue eyes rather than green.

Oddly, Piscean animals are apt to have hind legs longer than the front, giving them a somewhat rabbity appearance — made even more so by a stubby tail.

Pisces is, of course, a water sign and your Piscean cat, unlike other felines, may actually like water and be a good swimmer. I also notice that my Jacob drinks a great deal of water and make a point to replenish his supply daily, so he can enjoy himself. He was one of those cats who just came by and took up with us, so it was fun for me to identify his sign.

CHAPTER 2

CAPRICORN, OR GETTING YOUR GOAT

Few things have rewarded us more generously than our goats:
with milk, affection, entertainment, and fertilizer.

O ur first goat was Maggie, a hardworking, unpretentious gal of rather uncertain ancestry, probably mostly Toggenburg. We had lots to learn about goats, and she patiently (well, most of the time!) permitted us to practice on her.

Many people begin a goat-keeping venture because the illness of a family member necessitates a substitute for cow's milk. None of us had this need, but we did have two small children and we had heard much about the healthful properties of goat milk.

We found the milk to be everything we had hoped for, of fine flavor and high quality. And it is absolutely true that goat's milk is more easily digested than cow's milk because, among other attributes, it has smaller, finer, and more readily assimilated fat globules. Thus, it is also more nourishing, for we are nourished *not by what we swallow but by what we digest.*

GOATS NEED COMPANY

Our Maggie, who had freshened (began lactating) shortly before we bought her, provided us immediately with 3 quarts of milk a day, endless entertainment for the youngsters and ourselves, and frustration because of her frequent noisy blatting.

We couldn't figure this out, being new at goat-keeping. Maggie was well fed, comfortably housed, milked right on the minute, and she had a nice exercise yard with a teeter-totter and a series of sturdy boxes to climb on. She was apparently in tiptop health and ate like her feed was going out of style.

In desperation we finally went back to the old goat rancher from whose herd we had taken Maggie. "Figgered you'd come back," he said. "Maggie noisy?"

"Yes," we said. "What's wrong?"

"Nothing," he said calmly. "She's just lonesome."

"Why didn't you tell us?" asked Carl.

"Well," drawled the rancher, "at the time, I wasn't sure you'd believe me. Thought you might figger I was just trying to sell you two goats."

"Is that what it will take to quiet her down?" I put in. "Another goat for company?"

"Yes," he said. "Besides, you ought to have two goats anyway, so you'll have a continuous milk supply. Breed one at the beginning of the season and one at the end."

So we picked out Pet, a purebred Saanen who had just been bred to a good sire, put her in the back of our pickup, and brought her home.

From the moment Pet stepped into the goat yard, Maggie quieted down. They batted heads for a few days to determine supremacy and then proceeded to live together in complete compatibility.

It was shortly after Pet joined the family that I discovered another use for the milk goats. They became my nursemaids. Our children were brought up to treat all animals with kindness and they never teased any animal on our small "farm." Gradually, as rapport became established between the goats and the children, I came to trust them (the goats, that is).

The vegetable garden was adjacent to the goat yard, and one day I left the children with the goats while I worked. They were happy and entertained and I was free to concentrate on weeding. I lost all sense of time. Suddenly I realized that things were very quiet and then remembered that I had left the children with the goats. My heart skipped some beats. I looked into the goat run and the children were not there.

I rushed through the gate and into the stable expecting I knew not what, only to find the children fast asleep on the fresh, clean hay with Pet and Maggie standing by, nibbling placidly on wisps of alfalfa. As I entered the door, Pet stepped carefully over my sleeping son and lay down in her stall. After that, both youngsters often took their afternoon naps in the goat stable, sometimes lying with their heads pillowed on the resting goats, while I busied myself about the garden and other tasks. They were safe there.

MAGGIE DRIES UP

The time came at last when Maggie started to dry up and we knew that she would need to be bred again. Gestation is the time between impregnation and the birth of the young. In goats, this period is approximately 151 days.

We wanted to breed her so that she would freshen in a good sign. We determined the best time by subtracting from the fruitful dates the number of days given for gestation. The fruitful signs are the feminine water signs of Cancer, Scorpio, and Pisces (with Cancer, our choice, the most fruitful). Young born during the fruitful signs are usually more healthy, mature faster, make good breeding stock, and, if they are milk animals, produce better.

Animals born during the semi-fruitful earth signs of Taurus and Capricorn will be strong and healthy and mature rapidly, but will produce leaner meat. If you are breeding animals for showing or racing, choose Libra.

At just about this time Pet freshened, presenting us with a beautiful, snow-white Saanen buck kid. He proved to be friendly, mischievous, and utterly delightful.

A few weeks later we had a visit from a local doctor. He had a patient, scarcely more than an infant, with a digestive problem, and he wanted urgently to prescribe goat's milk from a clean, sanitary source.

Bouncy Snowdrop, as we had named him, wasn't quite yet at weaning age and he was a valuable little fellow. To take him away

Capricorn is the best sign for weaning young.

from his mother now wasn't what we had planned, yet to turn down the doctor didn't seem right either. A few days later, with some misgivings, we moved our fine, healthy little buck kid into the poultry yard with our white Rock hens. We had meant in any case to take Snowdrop from his mother when the moon was in Capricorn. We just did it a little sooner.

Though we consider Capricorn the best sign for weaning, success may also be achieved in the signs of Sagittarius, Aquarius, and Pisces. Whatever sign you choose, it is important that the mother nurse her youngster in a fruitful period for the last time. If Venus is well marked at this time, it will also help speed matters along.

CHANGE THE KID'S FEED GRADUALLY

We knew, of course, that whole goat's milk is the best feed for kids, but we also knew that dry skim milk from goats or cows could be used instead. We changed Snowdrop's rations gradually, mixing dry skim cow's milk and water to give the same milk solids as natural skim milk. Snowdrop, by this time, was drinking water from a pan, so we decided not to put him on a bottle but rather to exercise a little patience and teach him to take his milk from a pan as well. In this we were successful.

You can, if you wish, substitute a high-protein calf starter for part of the milk. But keep the calf starter and a small amount of high-quality alfalfa hay before the kid at all times, even if the kid won't eat much until it is about 3 weeks old. If the kid eats grain and hay well at the end of the eighth week, you may discontinue the milk feeding.

All kids start nibbling leafy hay when they are just a few days old. When they are 2 or 3 weeks old they start drinking water by themselves from a pan or pail, even if they are nursed or bottle-fed. And, if green plants are available, kids start early to eat leaves.

Mother and son adapted quickly to the new routine. Naturally, they were a little noisy for a day or two, but they could see each other and both settled down rapidly.

Snowdrop grew out well, and in time was duly presented to our county agent to be raised as a sire so that the rather scrubby stock in our area could be upgraded. Our first "patient" improved in health and the digestive disturbance cleared up. Eventually, things got back into the usual pleasant, customary routine.

Goat-keeping, like many other endeavors, is a very simple operation if you know what you're doing. It is particularly easy for women and older children. Goats may be kept, and kept well, in a small space. If you have a love of animals and a little patience, all the rest will fall into line.

BREEDS OF GOATS

The Saanen, one of the leading breeds, takes its name from the Saanen Valley of Switzerland. It is white or slightly cream in color. It produces the greatest quantity of milk. A mature doe weighs about 135 pounds. We always found our Saanen does to be very quiet.

The Toggenberg is also a leading Swiss breed. It is brown with a light stripe down each side of the face. The legs are white below the knees and there is a white triangle on either side of the tail. The Toggenberg, often referred to as the Guernsey of the goat family, produces rich milk in good quantity. A mature doe weighs from 115 to 150 pounds.

The Nubian is a native of upper Egypt, Nubia, and Ethiopia. It has a short, sleek coat; long, drooping ears; and a Roman nose. Colors range from red to tan or black, with or without white. Nubians are one of the best breeds for high butterfat production. A mature doe weighs 130 pounds and up.

The French Alpine was developed in the mountains of France and for generations has been selected for heavy milk production. The colors vary from white and gray to brown, black, and red, and show shadings and combinations of these colors on the same animal. Its outline is angular. Like the Toggenberg, it produces rich milk in quantity. The average doe should weigh not less than 125 pounds.

GETTING YOUR GOAT

When buying a goat, consider (1) the record of its production, (2) the production of its ancestors, and (3) physical appearance.

When buying an older animal, be sure to ask for a record of the animal's milk production and the production of its offspring. If the goat is young and has not yet produced milk, ask to see the records of the dam.

Because goats are dairy animals, they should have dairy characteristics. Notice if the doe has a feminine head, thin neck, sharp withers, a well-defined spine or backbone and hips, thin thighs, and rather fine bones. The skin should be thin and fine over the ribs. Look for a wide spring of rib and roomy barrel. Looking at the udder is important as well. It should be large when full of milk and very much smaller when empty. A large udder, however, does not always mean a high milk yield. If your goat is to run on brushy land, note whether the udder is high and well carried; a low-hanging udder will be subject to cuts and scratches. (This condition is more common in an older animal.)

MILKING YOUR GOAT

Equipment for a backyard dairy for two goats can be very simple. Basically you need only a good shelter, at least a small exercise yard, and a milking stand. The milking stand can be, like ours, of the folding type that attaches to a wall. Most garages will easily accommodate such a stand. (Even a stand is not absolutely necessary, but most people find it convenient.)

It is desirable to have a place for milking separate from the main goat barn. This prevents the milk from absorbing any odors in the stable. It is better for the same person always to do the milking, but there were times when Carl had to be away from home, so I did learn how.

As young does may initially object a bit to being milked, a stanchion arrangement is an excellent method of handling them. At first it is best to give the does a little grain feed in the box attached to the stanchion. Does soon become accustomed to being milked and will jump up on the stand and put their heads through the stanchion without assistance.

The doe's udder should always be either washed or wiped thoroughly before you milk her. Ordinarily a damp cloth is sufficient to remove all foreign material. Carl always kept the hair clipped near the udder of our milking does as well.

There are two systems of milking goats: from the side, as cows are milked, and from the rear. The latter method is largely European and is seldom used in the United States because the milk may be contaminated

For simplicity, milk your goat from the side at a stanchion.

by dirt and droppings. Commercial dairies usually make milking arrangements to conform to local health regulations. The individual owner may decide which method is more convenient.

MILKING METHODS

There are also two ways of drawing milk from the udder. The first is to press the teat in the hand, as is usually done in milking cows. This can be adopted when the teats are big enough to grasp.

The second method, known as stripping, is necessary only for goats with small teats or for goats in their first lactation, before the teats are fully developed. In stripping, the teat is grasped between the first finger and the thumb close to the udder and drawn down the entire length. Sufficient pressure causes the milk to flow freely.

Don't save the first milk drawn; the openings in the teats may be partially filled with foreign matter, which will be removed after a little milk has been drawn.

Goats are likely to hold back some milk. To prevent this, massage the udder and gently bump it up and down (as a suckling kid would) after each milking.

For a short time after kidding it may be necessary to milk a heavy producer three times a day, but twice is sufficient for most does. The period between milkings should be divided as evenly as possible.

Milk should not be used for human consumption until the fourth or fifth day after the doe gives birth. Some authorities recommend waiting a little longer, but this is not necessary if everything is normal.

Regularity in milking is important, and kindness and gentleness are, it goes without saying, essential.

CARING FOR THE MILK

Keep all utensils clean. After washing, our practice has always been to sun them well. A sanitary stainless-steel milking pail with a detachable hood is good to have but a small pail that readily scrubs clean will do. Pails are usually of 4-quart capacity.

As soon as the milk is drawn, weigh, measure, and cool it. The weighing is necessary to determine accurately how much a doe produces. Milk records are especially valuable to the breeder in selling stock as well as in selecting breeding animals.

The milk should always be thoroughly strained to remove any foreign matter. The best method is to use commercial filters or strainers, but you can use a layer of sterilized absorbent cotton between two pieces of material, or pass the milk through several thicknesses of clean cheesecloth.

To check the growth of bacteria, cool the milk to a temperature of 40°F as soon after milking as possible by placing the cans in a tank containing cold water. In a home dairy, where only a few quarts must be cared for, simply pour the strained milk into clean quart fruit jars and chill in the refrigerator.

Tuberculosis does not exist among goats. Since we always kept only healthy animals, we never felt the need for pasteurization of the milk, but this is required by public health authorities in some localities where goat milk is sold for human consumption. Studies have shown that with pasteurization, the solubility of calcium and phosphorus is slightly increased and the curd tension is reduced. This process improves the keeping quality more than the flavor of fresh goat's milk. Pasteurization by holding the milk at not lower than 142°F for 30 minutes caused a decrease of from 33 to 45 percent in the content of ascorbic acid, or vitamin C.

LACTATION PERIOD

The lactation period, the time during which a doe produces milk, varies according to breed and type of goat. It may range all the way from 3 to 10 months or even longer. Eight to 10 months is considered satisfactory. Certain conditions, such as the breed, individuality, health, feed, and regularity and thoroughness of milking, may influence lactation. We found that purebred does, of any of the leading breeds, will generally milk longer than the so-called common, or American, type.

The health of the does while giving milk is especially important. When does are out of condition, their milk yield frequently shrinks and they may have to be dried up. Proper feed, given with regularity, tends to extend the lactation period by stimulating the production and causing a more uniform flow. Milk regularly and thoroughly for long lactation.

A doe that produces 3 pints a day is considered only a fair milker; the production of 2 quarts is good, and 3 quarts is considered excellent, provided the lactation is maintained for from 7 to 10 months. Good does should produce from 8 to 15 times their weight in milk in a lactation period.

Goat's milk is nearly always pure white. Properly produced and handled, it should have no disagreeable odor or flavor. Bucks should never be permitted to run with the does. In fact, it is not advisable, and usually unnecessary, for anyone keeping only a few does to keep a buck at all. Contact your local county agent for help in finding a sire when it's time to breed. In most communities now it is possible to find a ranch or farm that provides this necessary service.

FEEDING YOUR GOAT

Though goats are capable of eating almost anything, they do much better on a balanced diet of grass, alfalfa hay, and root vegetables, supplemented with dairy pellets for the milkers. Given the chance, a goat heads for the nearest branch of leaves, much as a deer will, and they also like weeds and shrubs.

A fairly good goat, giving 3 to 4 quarts of milk a day, should be adequate for the average family. To produce this she will need about 4 pounds of hay and 2 pounds of grain, plus surplus vegetables and carrot and beet tops from your own garden, undamaged by sprays and poisons. Even now, keeping a goat is relatively inexpensive.

All that nonsense about goats eating paper and tin cans is just not true. A milk goat is discriminating in her feeding habits. Goats, being natural browsers, like to reach upward for their food; in a pasture they like nothing better than to nibble on young trees and may be relied on to clean off the lower limbs and eat out the underbrush. They do this so well that they are sometimes kept on large estates just for this purpose. However, if the underbrush is heavy, a goat of the Angora type will be a more practical choice than a milk goat.

If you have land with a reasonable amount of vegetation on it, several goats can live in an area that would not afford sufficient sustenance for one cow. They will even eat thorny shrubs, such as a rosebushes, with great relish and seemingly with no pain or inconvenience. Save all your rose and other nonpoisonous shrub trimmings for them and let them enjoy.

Some plants, however, do not make good fodder. Cherry, laurel, yew, buttercup, and wild larkspur are toxic, and some firs and pines are too rough for a goat's relatively tender mouth.

To suit goats' habits and to prevent waste, Carl constructed a large-mesh wire cradle in which to hold alfalfa or peanut hay. Immediately underneath this he placed a large, elevated wooden box to catch leaves and bits of broken stems. (We found that once hay fell on the floor of her stall, no self-respecting goat would touch it.)

OTHER GOAT FEEDS

Goats do well on abundant sweet water, iodine-rich foods, and foods rich in aromatic oils. In hot climates, particularly, they require access to rock salt. Their natural food is leafy and woody rather than grassy, and natural woodland grazing contributes much to total health.

Crops especially beneficial to goats are oats, planted with vetch for cutting whole, and barley. Barley is sometimes called goat cereal. Stretches of wild barley indicate a lime-rich earth, and for goats it is both a valued food and a medicine. Plant barley in the first or second quarter in Cancer, Scorpio, Pisces, Libra, or Capricorn. If you do not wish to feed your goats on a prepared chow they will do well on a tasteful blend of barley, oats, corn, and soybean meal, all bound by molasses.

Other crops good to grow for goats are alfalfa, sunflower (for both plant and seed heads), linseed, and corn (including the cobs). Flaked bran and rolled oats, as well as dried beet pulp, are excellent feeds.

During the winter months, goats relish silage, especially when mixed with molasses. This makes a welcome and nutritious change from dry hay.

A BARLEY BREW

Barley water has long proved a cure for ailments of the kidneys and bladder and is also a famed colic remedy for infant animals. Pour water, just off the boil, over an equal amount of crushed barley grains. Allow to stand overnight, then add an equal amount of tepid water. To each cupful of barley water add 1 teaspoon of pure lemon juice and 1 teaspoon of pure honey. Strain through cheesecloth, being sure to wring out the barley until it is very dry. Give as a drench in fever ailments and for disorders of the kidneys and bladder.

GOATS AND HERBS

I have long been convinced that one of the reasons goat milk is so helpful is the goat's natural tendency to seek out and devour many medicinal herbs. This can be true, of course, only of the free-ranging animal.

✩ ✩ ✩ ☽   FEEDING GUIDE   ☽ ✩ ✩ ✩

AGE	FEED	AMOUNT EACH DAY
Birth to 3 days	Colostrum	All the kid wants
3 days to 3 weeks	Whole milk (cow or goat)	2 to 3 pints
	Water, salt	All the kid wants
3 weeks to 4 months	Whole milk	2 to 3 pints, up to 8 weeks
	Creep feed [1]	All the kid will eat, up to 1 pound per day
	Alfalfa hay [2]	All the kid will eat
	Water, salt	All the kid wants
4 months to freshening	Grain mixture [4]	Up to 1 pound of high protein feed
	Alfalfa hay or pasture [2]	All the doe will eat
	Water, salt	All the doe wants
Dry pregnant	Grain mixture [4]	Up to 1 pound mix for a dry animal
	Alfalfa hay or pasture [2]	All the doe will eat
Milking doe	Grain mixture [3]	Minimum of 1 pound up to 2 quarts of milk per day. Add 1 pound grain mixture for each additional 2 quarts milk.
	Alfalfa hay [2]	All the doe will eat
	Water, salt	All the doe wants

[1] Creep feed may be a commercially mixed milk supplement or calf starter.
[2] Alfalfa hay of very high quality, fine stemmed, leafy, and green.
[3] Suggested grain mixtures: For a lactating doe, 55 pounds of barley or oats, 15 pounds beet pulp, 20 pounds wheat, mixed feed or mill run, 10 pounds linseed, cottonseed, or soybean oil meal.
[4] For a growing or dry doe, 15 pounds beet pulp, 50 pounds barley or oats, 15 pounds wheat, mixed feed or mill run, 20 pounds linseed, cottonseed, or soybean meal.
Note: If you use commercial feed for your dairy goat, use it according to the animal's stage of growth — growing, drying, or lactating.

Dandelions, blood cleansing and tonic, are avidly grazed by goats. Goats love the foliage of **elm trees.** They take shelter under these trees during storms as well, trusting the lightning-resistant powers that the trees have been credited with since olden times.

Goats love **honeysuckle.** In fact the French name for this herb, *chèvrefeuille,* translates as "goat's foliage." The whole plant is medicinal. **Lavender** is eagerly eaten, imparting a sweet flavor to their milk and cheese, and also aids the keeping qualities of these products.

Marjoram gives a sweet taste to goat milk. This aromatic plant is tonic and purifies the blood. *Parsley* improves milk yield and, high in iron and copper, enriches the blood. Goats love the foliage of *roses* as well as the fruits and flowers. All parts of the plant contain healthful vitamin C.

Rosemary imparts a fine fragrance to their milk, as well as having tonic properties. *Sage* makes their milk refreshingly tonic and increases the yield. *Sow thistle,* rich in minerals, and *sweet cicely,* aromatic and tonic, both increase milk yield and are beloved by goats. Such is the case with the tonic and antiseptic *thyme.*

One of the best herbs for goats is a seaweed such as *kelp.* This is particularly good for kids. Have you ever noticed that a kid will chew on wood for comfort as its teeth are developing? To keep the kid from doing this, offer it kelp and watch its happy acceptance. Probably only a goat or a dentist really knows why this works, but my guess is that the iodine in the kelp acts as an antiseptic.

Powdered kelp also aids in their glandular development. And if warts in your older goats are a problem, increase the iodine in their diet by feeding them seaweed. *Cleavers* (an annual bedstraw) is also helpful and, if possible, give them *watercress,* naturally sulfur-rich, and some *cabbage.*

These are only a few of the possibilities. Many other herbs may be given at times, or used externally for specific reasons. See Chapter 13, "Medicinal Herbs for Animals," for suggestions on herbs you can grow yourself if your land, or surrounding fields, does not yield them naturally.

GOAT MANURE

Goats, even as they produce milk, are also converting feed into high-grade manure. The goat's digestive system is such that seldom do weed seeds pass through undigested, so unwanted volunteers will not spring up in your vegetable patch. Goat manure has about the same composition as sheep manure. Apart from trace minerals and similar ingredients, goat manure contains about 64 percent water, 1.44 percent nitrogen, .22 percent phosphorus, and 1 percent potassium.

Goat manure, being dry and in pellets, is clean and practically odorless. It is excellent for composting.

BREEDING DOES

To maintain milk production over a period of years, it is necessary to breed your doe annually. Milk goats are good breeders. A mature doe will often have two kids at one time, but she may even have three or four.

Goats tend to be seasonal breeders, usually from late August through March. If you have two does, breed one at the beginning of the season and one at the end.

Does are generally bred for the first time at 12 to 15 months of age if they are well grown. During the breeding season, the buck has a strong odor, and should be kept in a separate pen at all times. The female goat is absolutely odorless.

Does usually remain in heat 1 or 2 days. The period between heats varies but is generally from 17 to 21 days. A doe will freshen about 5 months after the day of service. This ranges from 145 to 155 days; the average is 151 days. (Use the average in figuring the birth date.)

It is considered good management to have the doe freshen once each year. Allow her a dry period of six to eight weeks. To dry your doe, switch to dry feed and cut out concentrates. Do not milk her for seven days. Her udder will fill up, of course. This pressure turns the doe's system away from milk producing and dries her up. At the end of a week, milk her out again.

AT KIDDING TIME

Shortly before your doe is due to freshen, clip around her udder, hindquarters, and tail for greater cleanliness during the birth. Give her a quiet kidding stall and clean bedding. Do not tie the doe. Do not leave cold water where she can drink it after kidding. Because kids are often born when the doe is standing, don't leave a water bucket where she might drop one into it.

A few days before she is due, cut down her grain feeding. Substitute laxative feeds, such as bran or beet pulp. You will know when she is due by these signs: rising tailbone, loose to the touch, with sharp hollows on either side; rapid cud chewing; restlessness and pawing at bedding; low, plaintive bleating; rapidly filling udder, turning pink and shiny just before kidding; and a mucous discharge from the vulva.

We made it a practice to stay with our does at kidding time to ensure a smooth delivery. If your doe seems to be experiencing more than ordinary difficulty, call in a veterinarian to check her.

It was always my custom, also, to prepare a drink of bran mash mixed with warm water to give a doe shortly after kidding. She loved this, and I always let her have it for several days afterward as well.

CASTRATION

If you have a male goat that will not be used for breeding, castrate him when he is 7 to 14 days old. Do this on a bright, dry day, preferably when the moon is in Capricorn (knees), Aquarius (calves, ankles), or Pisces (feet). Avoid a time when the moon is in Virgo, Libra, Scorpio, or Sagittarius. Try to select a date when the moon is within one week before or after the new moon.

A veterinarian can perform this operation, or you may do it yourself. Use a clean, disinfected knife to cut off one-third of the lower part of the scrotum, or bag. Then force the testicles out and hold them with a firm grip, pulling them out with the attached cords. Do this with a steady pull; do not jerk. Cut or crush the cords and treat the wound with some standard disinfectant.

SLAUGHTERING FOR FOOD

In many parts of the world, goat meat is considered a delicacy. Even here, in the United States, a great many goats of the milk type, especially kids, are consumed annually. Sometimes goat meat is sold as lamb. Kids are usually slaughtered from 8 to 12 weeks of age. The flesh is palatable and has a flavor suggesting lamb, but remember that goats do not fatten and carry flesh as sheep do. You may hear the meat of goats referred to as chevon.

The best time for slaughtering is the first 3 days after the full moon. At this time the meat will be tender and have a fine flavor; as well as better keeping qualities. Do not slaughter in the sign of Leo.

BUTTER

Goat butter can be made from goat's milk, but seldom is. The cream rises very slowly and only a portion of it ever reaches the top. However, practically all the butterfat can be obtained with a separator. The butter is white but may be artificially colored (I dislike artificial anything) if your family finds "yellow" more appealing. Colored, it will resemble butter from cows but with a different texture. Use it for the table or for cooking.

CHEESE

Several kinds of cheese, known under various names, are made from goat's milk. Cheese may be entirely of goat's milk or contain one-fourth to one-third cow's milk. Many think a mixture improves the quality. The process is simple, and no unusual equipment other than a few special forms and a curing room (where the temperature may be kept at 60°F) is needed.

The fresh milk is set with commercial liquid rennet for about 45 or 50 minutes at a temperature of from 86° to 90°F. It is often advantageous to add 1 percent of starter. Rennet is diluted about 20 times in cold water and added at the rate of about 25 drops to 10 pounds of milk.

After a thin film of whey has collected on the firm, coagulated milk, cut it with a cheese knife into pieces about the size of a walnut. After the curd has remained in the whey for 5 minutes, stir gently for 5 minutes, and then pour into forms using a cup or a long-handled dipper.

These forms are made of 3X tin and are 4½ inches in diameter by 5 inches high. Each form has five rows of holes, the holes being about 1 inch apart and ⅛ inch in diameter.

The curd remains in the forms undisturbed until it acquires a consistency that will allow turning. After the curd has stood from 24 to 36 hours at a temperature of 70°F, apply natural sea salt to the surfaces. Leave the cheese on a draining board for about 24 hours. Then place the cheese on plain boards and carry them to the curing room, which should have a temperature of 60°F and high humidity.

A blue mold will appear on the cheese and should be brushed off with a moistened cloth. The next growth, slimy and reddish, seems to be necessary to bring about the proper ripening changes. It will cover the cheese. The curd, at first sour, becomes gradually less so, finally developing a sweet and pleasant flavor. When the acidity has disappeared, the cheese is ready for wrapping.

Wrap the cheese in parchment paper alone, or in parchment paper and aluminum foil (which prevents drying, promotes ripening, and gives the package an attractive appearance). Then put the cheese in regular Camembert boxes. Allow 5 or 6 weeks for ripening. The cheese should then have a fine, white color and an agreeable flavor.

DEHORNING KIDS

Some kids are born hornless. If your goat starts to develop horn buds, remove them when the moon is not in Aries, Taurus, or Pisces. The best time is one week before or after the new moon.

Caustic sticks of soda or potash may be used, but be very careful in handling them. They can injure the skin or be harmful to other goats that come in contact with the treated animal.

The best results, I believe, are achieved with a disbudding iron from a goat dairy supply firm. It is easier on both you and the kid than other methods.

Heat the iron so that at least 2 inches are *cherry red.* At a lower temperature, the procedure takes longer and is more exhausting to the kid. Save time by heating two irons, so a fresh one is ready for the second horn bud. For small doe kids, a ⅞-inch (diameter) iron is adequate; for large does and especially buck kids, a 1-inch iron is better.

Center the iron on the horn buds, applying with a circular motion and light pressure. Do this for about 5 to 10 seconds or a bit more, depending on the size and development of the horn buds.

When the iron has burned enough, the clean skull will show. Apply unguent or carbolated vasoline to each disk immediately after disbudding.

TRIMMING HOOVES

In their natural environment, which is high and rocky terrain, goats' hooves are worn down by normal wear and tear. However, most goats kept for a small home dairy do not have this advantage. To ensure your goat's good health, properly trimmed hooves are a must. Trim hooves when the moon is in Gemini or Leo, in the third or fourth quarter, with Saturn marked adverse.

Untrimmed or poorly trimmed hooves can cause serious lameness. The more often you trim them, the less you will have to cut off, but when you trim under the right sign, they will grow more slowly. Check the hooves at least once a month.

Use either a small hand pruner or a sharp knife with a blade that will lock in the open position. (A sharp knife is preferable to get a more level floor on the hoof.) Trim the bottom of the hoof so that it is parallel with the top. Always cut from heel to toe. The well-trimmed hoof does not have an overlapping wall, and the hoof floor is level and clean.

If the hooves are well cared for, you won't need to trim much of the pad, if any. Occasionally it may be necessary to trim some of the heel in order to get the bottom level. If the pad must be trimmed, do so in thin slices. Stop when the pad turns a pinkish color. You may draw blood if you go too deep.

Stand on the right side of the goat when trimming the front feet. Whenever possible, keep the animal against a fence or wall to prevent excessive movement. To work on the opposite hoof, reach across the animal and brace it against your body.

Working on only one toe at a time, remove the outer wall with the first cut. Next, even out the heel and pad to make the hoof floor level. After finishing the first toe, begin on the other, taking care to trim both toes so that when the foot is placed on the ground, one toe is not longer than the other.

When trimming the back hoof, stand to the rear, bring the goat's leg through your legs, and brace it against your knee. Proceed as you did for trimming the front feet. Oiling the hooves is good practice. Use pure linseed oil in which some rosemary has been steeped for several weeks.

Foot rot is a disease caused by an infection that destroys tissue. Lameness is a symptom, and the animal usually shows intense pain when the infected part is touched. To treat, carefully trim away the rotten part with a sharp knife, then paint with an antiseptic. A number of very good ointments especially prepared to cope with foot rot are available. Or try a steeped tea made from equal parts lemon balm, chervil, and chamomile, applying this as a compress. Keep the compress on for about 20 minutes and then dry it off. Repeat this process if necessary.

Keeping goats in dry pastures and clean, dry barns will go a long way toward eliminating the worry of foot rot. Clean away any broken glass and tin cans, too, to prevent injury.

GUARDING AGAINST DISEASE

As in all things, prevention is better than cure. The free-roaming goat in its natural habitat is almost never subject to the ills of the confined animal. Goats like plenty of direct sunlight but they also need shelter from the blaze in hot weather, and may even suffer sunstroke if this is not provided by a stable or the shade of trees. In nature, goats live on high ground; they do not thrive in low, wet pastures.

To get rid of *lice and ticks,* dip goats in a strong solution of derris, powdered if possible. A little eucalyptus oil in the dip will make it more effective. Camphorated oil is also good to add.

Sores caused by parasites should be treated with an application of garlic leaves or roots. Garlic given internally will assist in disinfecting the bloodstream and help to repel skin parasites.

For *worm* prevention, cleanliness is of the utmost importance. It will do little good to dose for worms if you continue to keep the animals on a worm-infested pasture. Dose the animals on the old land and then move them to another pasture. Cleanse the old pasture by liming, planting with garlic, or growing a heavy crop of mustard and then tilling it into the soil.

Garlic with molasses is one of the best treatments for worms, as it possesses great powers of pulmonary penetration. The lungs of goats (of sheep, too) can be an area for worm infestation and are difficult to reach when chemicals are used. Worms are most active when the moon is full. Treat at this time.

Garlic is also a good preventative. In the wild, goats will eat it avidly, grazing every reachable leaf. Areas planted to mustard make good grazing for goats, and mustard seed, either the black or white variety, is a safe vermifuge for kids. Two ounces per kid, given in milk, is the usual dosage.

Food additions such as carrot, pumpkin seeds, nasturtium, papaya, melon, grapes, and even raw desiccated coconut will assist in worm removal. Raw grated radish, both common and horseradish, and turnips are good too.

Mastitis, inflammation of the udder, will seldom occur if proper sanitation is observed. If the goat is with kid, allow the kid to suckle but keep the milk healthy by feeding the doe large garlic balls or garlic tablets once or twice a day. Garlic quickly enters the bloodstream and acts to purify.

Follow these practices to prevent mastitis:

- ✿ Keep barn floors and pens clean.
- ✿ Get rid of anything that may cause injury to the udder.
- ✿ Sterilize the teat cup and disinfect hands before and after milking each goat.
- ✿ Before milking, wash teats with a common antiseptic such as chlorine solution or any other relatively odorless antiseptic.
- ✿ Discover the disease early.

Goats make gentle nursemaids and produce fine milk and cheese.

CHAPTER 3

VENUS, THE "BEE STAR"

The Maya, who achieved outstanding success in astronomy and arithmetic, called Venus *Xux-eh,* the Bee Star.

Fossil bees found trapped in amber are believed to have lived 50 million years ago. Now these industrious and sometimes fearsome insects live in almost every part of the world, except near the North and South Poles. Of the 10,000 species of bees known, only honeybees make the honey and wax that are useful to humans, and they are the only insects to produce food that humans eat. Wax from the nests of bees makes such useful products as candles and lipsticks. Their delicious honey is enjoyed both for cooking and as a spread on bread. Moreover, honey contains definite medicinal properties. Bees and beekeepers are ruled by Mercury, Virgo, and Venus.

HIVING A SWARM

Beekeeping is a fascinating adventure, and my first experience with it was a most rewarding one. When Carl and I decided one spring that we would like to keep bees, we purchased the equipment necessary for one hive from the A. I. Root Company (see Resources), and Carl put together the "knocked down" pieces and stored them for the time being. This was fortunate; if we had not been prepared, we might have lost the swarm that appeared a week or so later.

Upon entering the garden one sunny spring morning I became aware of something unusual at the far end where our small orchard had been planted. I had never actually seen a swarm of bees before, but the movement on the bulging limb of a young apple told me something was going on, and I hastened to investigate. Sure enough, there hung a handsome swarm, the bees buzzing a little but not moving much. They were drunk on honey, having filled themselves, as is their custom, before leaving the old hive with an old queen to start a new colony.

I did know enough about bees to recognize that these were the gentle Italians that we preferred, as Carl had kept them before and considered them both easy to handle and good workers.

I rushed back to the house, wildly excited, and donned coveralls, the bee veil (slipped over an old hat), and gloves. I need not have done this, for those bees weren't going to sting anybody. (I still consider protective clothing advisable, though, even with the gentlest of bees.)

Not quite sure how to proceed, I went to the garage and picked up the hive. Pretty nervous about the whole thing, I carried the hive body down to the orchard, placed it immediately beneath the swarm, and gently shook the limb. The bees dropped in and began moving about their new home. I replaced the top and moved the hive to a stand we had prepared in a shady place. Apparently, in my blundering way, I did the right thing.

I was to learn later that my prompt action in moving the swarm probably was the reason the bees remained. The advice given to beekeepers is to hive a swarm quickly and then move it so that returning scouts cannot lead the swarm away. Sometimes bees travel several miles before deciding on a new home, clustering and breaking several times while moving around.

The eventual success of this first hive was due, I'm sure, to Carl's knowledge of good management, for successful it was. That summer we put on super after super (upper stories added to the hive), which the bees diligently filled. When it came time to "rob" the bees, there was plenty for us and ample stores to leave for their winter food.

IF YOU ARE THINKING ABOUT KEEPING BEES

Getting started with bees is not always so easy, or so inexpensive, but purchasing bees from a reliable source is usually not a great problem. Just about anyone who really wants to keep bees can do so. You will need:

- ✿ A modest investment in materials.
- ✿ A suitable location for beehives.
- ✿ Elementary knowledge of the habits of honeybees.

One of the best ways to get more information is to talk to a local beekeeper. Beekeepers, like gardeners, are always proud to show off a little. He or she may even (weather permitting) open the hive and handle the bees, showing you how to reduce the likelihood of being stung and how to get the honey out of the hive.

Your county agricultural agent can supply you with pamphlets or direct you to other information sources. You may even want to take a course in beekeeping at your state agricultural college.

Much can be learned by joining a beekeepers' organization; most states have at least one. The library is a source for books, magazines, and journals on beekeeping.

BEEKEEPING EQUIPMENT

Here is the basic equipment you'll need to start beekeeping:

- ✿ **Hive,** to house your bees.
- ✿ **Frames,** to support the honeycombs in which your bees will store honey and raise young bees.
- ✿ **Smoker,** to blow smoke into the hive. This pacifies the bees when you want to work with them.
- ✿ **Hive tool** with which to pry frames apart so you can examine the hive or harvest the honey.
- ✿ **Veil,** to protect your face and neck from bee stings.
- ✿ **Gloves,** to protect your hands.
- ✿ **Feeder,** to dispense sugar syrup until bees can produce their own food.

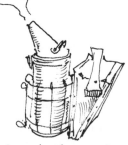

A smoker is a must for pacifying bees.

STRAINS OF BEES

Many races of bees can produce good amounts of honey, but some are so fierce that keeping them is not advisable. Our preference has always been for the gentle, hardworking Italian strain.

The Caucasian is another strain of bees widely kept. They are even more gentle than the Italians but the queens are dark and thus difficult to find in a cluster of bees. It is important to the beekeeper to be able to find the queen bee, for it may be necessary to replace her after a year or two if she doesn't lay enough eggs to keep the colony strong.

Another disadvantage of Caucasian bees is that they use an excessive amount of propolis in their hives. They collect this gummy substance from buds and injured tree parts, using it as a kind of "cement." Frames that become heavily propolized are difficult to remove.

Some specially developed hybrids, crosses between two or more bee strains, are more productive than standard strains. However, after a year or two, their offspring may bear little or no resemblance to the original hybrid bees. If you keep hybrids, replace your queen each year to ensure a uniformly strong colony.

THE COLONY LIFE

The honeybee is a social insect. This means that bees live together in a colony and depend on each other for survival.

Most of the bees in a colony are workers (that is, sterile females). Some are drones (males), whose only function is to mate with the queen. Usually there is one queen (fertile female) in the colony; after mating, she lays eggs that maintain or increase the colony's population.

Worker bees number from 1,000 to about 60,000, depending on the egg-laying ability of the colony queen, the space available in the hive for expansion, and the available or incoming food supply. During the greatest period of activity (the spring and summer months), they live about six weeks. They collect food and water for the entire colony, do the housework, and guard the hive against intruders. They also "air condition" the hive by fanning their wings to maintain a constant hive temperature and humidity.

Although worker bees do not mate, they may lay eggs if the colony loses its queen. However, their eggs will not keep up the colony population, because they develop only into drones.

The number of **drones** in a colony varies with the season. There may be none in winter, but several hundred during the summer. They are driven out of the hive in the fall, when worker bees can no longer collect food.

The **queen bee** normally flies from the hive when she is about a week old and mates in the air with one or more drones. When she returns to the hive, she begins to lay eggs, as many as 1,000 in a day. She puts each egg into a separate cell of the honeycomb.

A colony needs a new queen if the old one disappears or becomes feeble, or when the old queen and part of the colony decide to leave and build a new hive — in other words, **swarm.**

In some unknown way, the workers select a few larvae to become queens. These are fed only on **royal jelly** (a marvelous substance, creamy in appearance and rich in vitamins and proteins, formed by glands in the young worker bees). Many scientists believe the bees may add a special substance to the queen's royal jelly to make her grow faster and have a different appearance from the workers.

While this is going on, other workers build special cells for the queens to grow in. A queen cell somewhat resembles half a peanut shell hanging from the honeycomb. Somewhere between 5 and 6 days after hatching, the queen larva becomes a pupa. Sixteen days after the egg is laid, the young adult queen crawls out of the cell.

Three days after an egg is laid by the queen, it changes to a larva. Royal jelly is placed in the bottom of each cell, where it is believed to provide enough food for 3 days. When the larva is 3 days old, the workers begin feeding it a mixture of honey and pollen called **beebread.** Only a larva destined to become a queen is fed on royal jelly during its entire period of growth. It is the consumption of this royal food that makes a queen; the coarser food will make a worker.

As with the queen, worker bee "nursemaids" feed and care for the larva until it metamorphoses into a pupa. Then they seal the pupa into its honeycomb cell and leave it to finish developing.

Worker bee at
honeycomb

Twenty-one days from the day the egg was laid, an adult worker bee emerges and begins to work in the hive. Drones take 24 days to develop fully.

Whatever their origin, honeybees make use of the sun instinctively. A worker bee, upon returning to the hive, dances to tell the colony where nectar can be found after a source has been discovered.

GETTING STARTED

The best way to get started beekeeping is to buy a colony already established in a well-constructed hive that has honeycombs built into removable frames.

If you already have a hive, you can buy a package of 2 or 3 pounds of bees with a queen from another beekeeper or from a bee supplier. Be certain the bees you buy have a certificate of inspection to indicate that they are free of disease.

Another way to begin keeping honeybees is to do what I did: Capture a live swarm and establish it in your hive. Bees, drunk on honey, are usually gentle when swarming, but even so, wear gloves and veil and be careful. It is also sometimes possible to transfer a colony, with its combs, from a cave or tree to the hive. Capturing and transferring are probably best tried after you have worked enough with bees to feel relaxed around them.

Here's another pointer. Whenever you plan to work around your hives or handle bees in any way, it's a good idea to bathe first and put on fresh clothing; bees are extremely sensitive to odors. Perspiration in particular causes them to become excitable.

ESTABLISHING A NEW COLONY

The best time to establish a new honeybee colony is in spring. Fruit trees and flowers are in bloom then, and should supply the colony with sufficient nectar and pollen. And the bees will be beneficial in pollinating the fruit trees.

If you begin with a new swarm or package of bees, provide them with sugar syrup, that is, a mixture of half sugar and half water. Put the syrup in a feeder at the entrance to the beehive. The syrup keeps the bees from starving until they can make and store their own honey.

Lemon balm is particularly beloved of bees, so much so that some beekeepers rub the inside of a new hive with *Melissa offinialis* just before hiving a swarm. (If you do this, it is said that the bees will never leave the hive.)

PLACEMENT OF BEEHIVES

When your hive is stocked with a bee colony, put it where the bees are unlikely to be disturbed so that stinging is avoided.

If you live in a warm area, put the hive in the shade. In an area that has extensive periods of freezing or near-freezing temperatures, expose the hive to the sun and give it protection from prevailing winds, particularly in winter.

Always be sure that there is a constant supply of fresh, cool water nearby.

Blossoms from lime and linden trees are a great tonic for bees. Plant them in or near your apiary.

NECTAR INTO HONEY

Bees can't make honey without nectar, a sugary, liquid substance produced mostly by flowers. It is the raw material of honey and the bees' main source of food.

Sources of nectar for producing surplus honey vary from place to place. As a beekeeper you will be interested in learning which plants in your area are best for honey production. The "experts" say that it doesn't pay to grow honey plants for bees. Maybe not, but must the value of everything be figured in dollars and cents? And certainly you will at least want to know if there are sufficient honey- or nectar-producing plants in your locality to support bees before you decide to start an apiary.

In most parts of the country, bees work to full capacity only about twice during each summer; the rest of the time, nectar may be scarce. Planting something to bloom during their already busy times would be rather useless, but providing plants, trees, or shrubs that produce during a slack period could be very helpful.

Take willow honey, for instance. This is a dark, almost black honey with a bitter flavor (to many people, it tastes like medicine). Yet to bees it is very important. Many queens will not start to lay in the spring until new nectar and pollen are brought into the hive, and in many parts of the country willow is the earliest source of both. But since bees won't fly far in cool weather, the willow must be growing near the hive. All willows are easy to start; just push a branch into moist soil and it will root.

On the Menu

Bees greatly favor aromatic herbs such as lavender, hyssop, rosemary, catnip, thyme, santolina, dandelion, sage, goldenrod, bee balm, basil, marjoram, horsemint, and summer savory. They love aster, star thistle, alfalfa, buckwheat, cotton, sumac, and catclaw. Bitters such as southernwood, wormwood, and rue are treats. The mints are favored; they love all the rose family — wild roses most of all, but also blackberry and hawthorne. They adore members of the borage family: borage, alkanet, viper's bugloss, and anchusa. They are drawn to ling and bell heather, all kinds of *Scabiosa*, and, in the carnation family, clove pinks. Bees love the blossoms of such trees as tupelo, acacia, carob, olive, tulip tree, sourwood, oak, poplar, holly, mesquite, and palmetto, as well as fruit trees of all kinds, including citrus, and the blossoms of many weeds and grasses.

Maple honey has an excellent flavor, but the maple blooms so early that there are seldom enough mature bees to harvest all the nectar. Maple is crucial, though, for the security of the hive and may be a powerful force to start the queen laying.

Almost all fruits and berries depend on bees for pollination and often provide the first surplus, which may be a blend of several kinds of nectar with a goodly amount of dandelion. Think of the taste of pure apple-blossom honey, or blueberry or raspberry, where there are enough trees or bushes to make this possible. The creamy white, fragrant blossoms of black locust produce a beautiful champagne-colored honey, considered among the tastiest.

For something quicker to grow than trees there is the little white Dutch lawn clover. White-clover honey is the standard of excellence by which all other honeys are measured.

The color and flavor of honey depend on the kinds of plants from which bees collect nectar. Honey may be clear (nearly white), amber, reddish, or so deep brown as to be almost black. Its flavor ranges from mild to strong.

POISONOUS PLANTS

It may seem that bees can gather nectar and pollen from just about any plant or tree, but this is not entirely true. Certain plants and trees are toxic to bees.

One of these is the California buckeye. Bees poisoned by it become black and shiny, the result of hair loss. They may appear to be shaking, similar to the trembling that accompanies the last stage of paralysis. Queens affected usually produce eggs that do not hatch, but sometimes they do not lay eggs at all, or lay only drone eggs.

Other poisonous plants include black nightshade, death camas, dodder, jasmine, leatherwood or summer titi, locoweed, mountain laurel, seaside arrow grass, western false hellebore, and whorled milkweed.

When you do find toxic plants, there are several steps you can take. Depending on how numerous they are, you can either cut them down entirely or remove the blossoms. If plants are cut, and you wish to eliminate them permanently, do so in the fourth quarter in a barren sign such as Aries, Gemini, Leo, or Aquarius.

If cutting the plants (or trees) is not practical, relocate the hive during the period that the toxic plants are in bloom.

POLLEN

As worker bees go about their happy way, gathering nectar from flowers, tiny particles of pollen stick to their bodies (mostly on the legs) and are carried back to the hive. The bees store this pollen as beebread in cells of the honeycomb. Later it will be fed to young bees that are developing into workers and drones. (As previously mentioned, the few young larvae selected by the workers to become new queens are fed a special food, royal jelly, made by the workers in their own bodies.)

Pollen, then, is necessary for producing the new worker bees that become the new honey makers. An average-size bee colony uses about 100 pounds of pollen each year. That is why you need to locate your colonies near good sources of pollen, including many wildflowers, ornamentals, weeds, shrubs, and trees. Some especially good sources are aster, corn, dandelion, fruit blossoms, goldenrod, grasses, maple, oak, poplar, and willow.

HOW HONEY IS MADE

The nectar that bees collect is generally half to three-fourths water. After nectar is carried into the hive, the bees evaporate most of the water from it. While the water condenses, they change the nectar into honey. Then the bees seal the honey into cells of the honeycomb.

Beeswax begins as a liquid made by glands on the underside of a worker bee's abdomen. As it is produced, it hardens into tiny wax scales. Worker bees then use this wax to build the honeycomb.

Beekeepers often provide their bees with a honeycomb foundation made of sheets of beeswax. This foundation fits into hive frames and becomes the base of the honeycomb. It enables bees to speed up comb construction, and it provides a pattern for building a straight and easy-to-remove honeycomb.

MOVING A COLONY

When moving your bee colony, remember that it is important to orient the bees to the new location. If this is not done, unless you move the colony at least several miles, the bees will find their way back to their old location.

To move your bees only a few hundred yards, first take them several miles away and leave them for about a week. When they are oriented to the new location, move them to the site you originally decided on, and let them get oriented there.

Another way is to move the colony a few feet each day, until you have moved it to the location you want.

It is inadvisable to move bees during the heavy period of honey production. The honey already stored will add extra weight; new honeycomb may break loose; and you may so disturb your bees that you cause a slowdown in honey storage.

Night is the best time to move a colony. All the bees are then inside the hive. And if the weather is cold, you can completely close the entrance.

When the weather is unseasonably warm, and the colony is strong, do not seal the hive entrance. This may cause the bees to suffocate, even though they are sealed in for only an hour. Instead, cover the entrance and top of the hive with a fine screen.

Be sure to staple, crate, or tie the hive so that parts cannot shift during the move. Handle the hive with care, and avoid jolting.

MANAGING YOUR COLONY

As your bee colony grows, it will require more room. When the bees become too crowded, and there isn't enough room to expand the brood-rearing areas, they will swarm (fly off in large numbers, along with the queen, to form a new colony). You want to prevent this, of course: Losing a swarm of bees may leave the remaining colony too weak to store surplus honey.

To make more room for your bees, add extra boxes or stories of combs (supers) to the hive, or onto the supers already in place. One of the greatest improvements in modern bee culture is the use of a queen excluder, which prevents the queen from laying larvae in the upper stories. The use of the excluder will prevent difficulties later on, when you wish to take away frames of pure honey without larvae mixed in with it.

A standard beehive showing supers.

When the time comes for you to remove honey from the hive, do leave plenty for the bees. Remove only as much as you estimate to be more than they can use. And be sure there is at least 50 pounds of soft honey in the hive when winter begins; without it, your bees may starve before springtime.

Since one frame holds 3 to 5 pounds of honey, an average colony needs about 10 to 15 frames of honey to survive a winter.

HONEY PRODUCTION

Beekeepers usually measure honey production in pounds. The average yearly production of surplus honey from a colony is about 50 pounds. A well-managed hive, however, often produces several times that amount.

THAT YUMMY HONEY

Perhaps the most efficient way to get honey out of the comb is to uncap the honey cells with a warm knife (there are special knives made for this purpose), and spin the liquid honey out of the cells in a honey extractor. The honey is poured off, and the emptied comb is returned to the hive, ready to be refilled with honey.

With just one or two hives, though, it may not pay to buy an extractor. Perhaps you can rent or borrow one from a bee dealer, a neighbor, or a local beekeeper organization. However, you must consider that using borrowed equipment sometimes spreads bee diseases.

The least expensive, but also the least desirable, way to harvest liquid honey is to cut out the entire comb, squeeze the honey from it, and then strain the honey through a coarse cloth to remove wax particles. Although the bees cannot use the crushed comb again, you can melt it and sell the beeswax that you salvage.

Stainless-steel and glass utensils used in gathering and bottling organic honey are better sterilized by heat, rather than by antibiotics or other chemicals.

COMB HONEY

Some beekeepers produce comb honey by cutting out pieces of honeycomb. These are put in glass containers, and then liquid honey is poured around them.

Another way is to place small wooden boxes or "sections" in the top of the hive just before the honey flow begins. The bees will fill the sections quite neatly with honey, placing about a pound in each section. Remove the sections as soon as they are filled to avoid problems with honey dripping or leaking. Further handling and processing will also be unnecessary.

GRANULATED HONEY

Fresh honey has the best taste, whether in the comb or in liquid form. Some honeys granulate or become sugary even when fresh, however, and most will granulate sooner or later. Granulated honey is still good food; in fact, some people like it better than either liquid or comb honey. But if it's not to your liking, it may be liquefied by the method that follows.

1. Place jars of granulated honey in a container with enough water to reach to the level of honey in the jars.
2. Support the jars so that they do not rest directly on the bottom of the container and so water can circulate around them.
3. Heat gently, until granules have disappeared. (The time required for heating cannot be given exactly. It will vary, depending on the size of the jars of honey and the temperature to which you heat them. *Do not* heat water above 160°F; excessive heating darkens your honey and lowers its quality.)
4. Stir occasionally, to distribute heat evenly throughout the honey, and to determine when the granules have disappeared.

You can freeze honey, thus preserving the delicate balance of nutrients sometimes destroyed by careless heating. Let the honey set for at least 48 hours to clear up. Frozen honey will retain, for at least six years, its original taste, aroma, and color, along with all minerals, vitamins, and enzymes. It will eventually crystallize, but you can restore it to liquid by the same method as for granulated honey.

HONEY AS A FOOD

Small amounts of pantothenic acid, pyridoxine, and biotin are present in honey. Here are the minimum daily requirements (MDR) of thiamine, riboflavin, niacin, and ascorbic acid for an infant, and the amount present in 100 grams of honey:

HONEY	MDR FOR INFANT	PRESENT IN 100 G	PERCENT OF REQUIREMENT
Thiamine	0.25 mg	0.004 mg	2
Riboflavin	0.50 mg	0.028 mg	5
Niacin	4.0 mg	0.12 mg	3
Ascorbic acid	10.0 mg	4.0 mg	40

Among the mineral elements found in honey are iron, copper, sodium, potassium, manganese, calcium, magnesium, and phosphorus. All of these minerals are essential to the good nutrition of animals. Each is present in honey, though in some cases only in trace quantities.

Honey is a pleasant and readily available source of energy for young children, and acts as a mild, natural laxative. Combined with molasses, half and half, honey provides both vitamins and a good supply of iron.

WARNING

Despite its many benefits, honey can cause infant botulism if given to children younger than 12 months of age. Do not give honey to children under 1 year. Do not give unpasteurized honey to children under 2 years.

WHAT'S IN A STING?

As a beginning beekeeper, you certainly want to know what happens when a bee stings you: The bee's stinger is barbed and has a poison sac attached to it. When you are stung, the barb and sac usually tear out of the bee's body. Convulsive movements in the sting muscle then push the stinger deeper into your flesh and venom is pumped into the wound.

Remove the stinger immediately by scraping it off with your fingernail or a knife blade. Don't try to pull it out; this only forces more venom into your skin.

Stings are intensely painful when first inflicted, and the pain is usually followed by reddening and swelling near the sting. Normally the pain subsides after a few minutes, though the swelling may persist for a day or two. Do not scratch the wound. A paste of baking soda and cold cream will help to minimize the pain. The juice of the pimpernel also soothes the stings of bees and wasps.

Usually you'll develop a resistance or immunity to stings after you've been stung a few times. Some people, however, are allergic to stings and can develop a severe reaction to them. Consult an allergy specialist if you plan to work with bees.

There are those who think bee stings are beneficial. The belief that people, when stung by bees, are cured of rheumatic ailments has been handed down for generations. It has been reported that beekeepers who acquire some immunity to the effects of stings never suffer from arthritis, rheumatism, or gout.

TO AVOID STINGS

Smoke pacifies bees, so always use a smoker when working with them. But use just enough smoke to keep the bees from stinging you. The amount will vary, depending chiefly on the strain of bees and the weather.

Direct the smoke into the hive entrance before you disturb the bees. When you remove the hive cover or a super, apply smoke to the bees as you expose them.

Wear protective clothing. A veil should be over your head and face. I

Wear protective clothing and use a smoker when working with bees.

place it over a wide-brimmed hat, tucking the ends inside my coveralls, which should be light-colored and sealed at the ankles, wrists, and neck. Wear canvas gloves or gloves made of thin rubber.

Remember, too, that bees are more irritable in cool, cloudy weather than they are when it is warm and sunny.

Don't confuse your bees. They're very sensitive; don't let them think you're a flower. Avoid wearing perfume or scent of any kind or shirts or dresses with a flower design. It's also a good idea to take a bath and wear fresh clothing before working with bees.

DISEASES, MITES, AND PESTS

Although several diseases attack honeybees, none is dangerous to humans. Nevertheless, most states have laws to control bee diseases. In many states it is illegal to offer for sale bee colonies without a certificate indicating that they are disease-free. If you suspect the presence of disease, consult your county agricultural agent. Mites are a far greater problem than diseases, and our knowledge about mites is still evolving. Consult your county agricultural agent or your beekeepers' association for up-to-date information on mites in your area.

Pests are best controlled by keeping the colonies strong. This is always good beekeeping practice. It is also your best protection against wax moth larvae, a serious insect pest that invades unprotected honeycomb.

Ants can be troublesome, especially in the southern states. The best cure is to destroy their nests. Keep the grass down around the entrance to the hive, and see to it that nothing is left that will form a "bridge" to the opening. In constructing a beehive shed, add cups to all the posts used to support the structure. Fill the cups with coal tar, creosote, or crude petroleum. Ants dislike these intensely.

A BOOST FOR BEES, OR A QUICK CURE

A medication effective for bees suspected of illness is composed of 4 tablespoons pure honey, 2 pints water, and ½ cup finely chopped sage. Simmer water and sage, covered, keeping well below the boiling point. Let stand until lukewarm, then stir in the honey. Allow to cool completely, strain mixture, then pour into the hive feeding trough.

Note: Do not substitute white or brown sugar for honey. Brown sugar may cause scour.

LIBRA, FOR BEAUTIFUL BIRDS

For show stock, mate your feathered friends in Libra.

Mate songbirds in Taurus, which rules the throat.

Have you ever lifted an egg from its carton and felt the thin shell crumble while its pallid contents dripped slowly through your fingers? This is a far cry (or cackle) from the hard-shelled egg of the free-ranging hen, with its firm white and its proud, upstanding, golden yolk. And the taste of such eggs, vitamin rich as well as appealing to the eye, is entirely different from that of eggs laid by the caged hen.

I am not going to argue with the dollars and cents of modern, commercial egg production. What we are going to consider here is *quality*, for our own well-being and for that of our flocks. I further believe that with careful planning, the expense of keeping a small flock of poultry raised for both eggs and meat need not be great. And with careful management, there is

Always store eggs large end up to keep the yolks well centered.

another by-product of value: manure. Composted before being placed on your garden area, chicken manure will add tremendously to the fertility of your soil. This manure is classed as "hot"; a little will go a long way.

GETTING STARTED

First, think about your objectives. Since we hoped eventually to raise our own chicks, we felt that it was important to choose a breed to which maternity would not come as a complete surprise. Many of the breeds commonly sold today, even the heavy types, are quite incapable of reproducing themselves without the aid of incubators and brooders. In the desire for greater egg production, they've had all the "broodiness" bred out of them.

We wanted chickens of a breed that would give us meat as well as eggs, for we intended to buy "straight-run" (unsexed) chicks and knew they would average out about half cockerels. We were also looking ahead to the day when our hens would no longer be profitable egg producers and would become "bakers" or "stewers." After much deliberation we decided that, for us, the perfect answer would be white Plymouth Rocks. (Besides, I'm partial to their brown eggs, with their warm shades ranging from cream to buff.)

Other breeds that make good mothers are buff Rocks, light Brahmas, buff Cochins, black Jersey Giants, and Araucanas (which lay the interesting, many- colored eggs). However, if you are more interested in egg production than anything else, the modern white Leghorns are widely noted for their ability in this line.

Young cockerels of many breeds make delicious eating. If this is your prime objective, you might wish to raise hybrids such as the Rock-Cornish, developed by mating Cornish chickens with white Plymouth Rocks. Although large when fully mature, Rock-Cornish chickens are usually butchered when only 6 weeks old. At this age they are quite uniform in size, each bird weighing about 1¼ pounds and usually supplying enough meat for one person. They are marketed as Rock-Cornish hens, quite expensive when purchased in the supermarket, but considered a great delicacy.

SEXED OR STRAIGHT RUN?

Chicks are purchased from hatcheries either "sexed" or "straight run." Sexed means you can buy just females or males. When you buy straight run, you don't know what you're getting but the chicks are cheaper. I suggest buying not fewer than 25 or more than 50 if you are a beginning poultry-keeper.

BROODING EQUIPMENT

The brooding period is that part of the chick's life from the day it hatches until it is 8 to 10 weeks of age. The chicks must have a warm, dry, ratproof place to live, plenty of water, and all the feed they need. They must also have fresh air, so the brooder house should have at least one window that can be opened and closed as appropriate.

Your brooder, or heating device, should provide heat for the chicks as follows:

AGE OF CHICKS	TEMPERATURE
1 day to 1 week	95°F
1 week to 2 weeks	90°F
2 weeks to 3 weeks	85°F
3 weeks to 4 weeks	80°F
4 weeks to 5 weeks	75°F

After 6 weeks, temperature should be kept at 70°F as long as needed.

You may use a conventional brooder with a hover or an infrared heat lamp to keep your chicks warm and comfortable. The hover-type brooders can be heated by oil, gas, or electricity. They are generally preferred if you are starting more than 100 chicks. You will need a thermometer to measure the heat under the hover.

An infrared lamp is a good device to raise 100 chicks or fewer. The lamp should be made of Pyrex glass so it will not break easily. Hang it by an adjustable chain about 16 inches above the liner. You cannot measure heat accurately with a thermometer under a heat lamp, so you will have to watch your chicks to see that they are comfortable.

A brooder guard should be used for the first two or three days of brooding, whether you use a hover-type of brooder or a heat lamp. The brooder guard keeps the chicks from wandering away from the heat and dying. You can purchase a guard quite inexpensively, or you can make one from cardboard or tin. It should be about 12 inches high and long enough to go around the brooder. Leave at least 3 feet between the circular guard and the hover's edge.

LITTER FOR THE CHICKS

Place litter on the brooder house floor. Materials that can be used for litter are wood shavings, vermiculite, crushed corncobs, or, in a pinch, sand and chopped straw. Put newspaper, or some other rough paper, over the litter to stop the chicks from eating it for the first three days.

FEEDERS

You will need small, chick-size feeders to start with. As a rule, the following feeder space per chick is about right:

Age of Chick	Linear Inch per Chick
1 day to 3 weeks	1 inch per chick
3 to 6 weeks	2 inches per chick
6 to 10 weeks	3 inches per chick

To figure feeding space, count the length of both sides of the feeder. A feeder 24 inches long has 48 linear inches of feeder space and is large enough to take care of 50 chicks to 3 weeks old.

WATERERS

Provide two small founts (glass or plastic base with quart or half-gallon jars) or one 1-gallon fount for each 100 chicks for the first two weeks. Two small founts are better than a single large one because they can be set out in such a manner that the chicks will not have to move very far to drink.

PREPARING THE BROODER HOUSE

Before your chicks arrive, plan on a suitable place for them to live. You may already have a brooder house, or perhaps you must build one.

If you have one that has been used before, it must be cleaned. Sweep down the walls and ceiling to remove any accumulated dust and cobwebs. Then wash it down with warm water and a good detergent. After the dirt has been removed, wash it down again, adding a good disinfectant to the water.

Oil of eucalyptus spread about the area will cause parasites to leave. An effective mixture is 3 parts oil of cloves, 5 parts oil of bay, 6 parts oil of eucalyptus in 150 parts alcohol with 200 parts water.

Repair all cracks in the walls and ceiling of the brooder house and replace any broken window panes. Don't be tempted to substitute paper, cardboard, or old sacks.

Place litter on the floor and spread it evenly, then set up the brooder and feeders. Place the brooder in the middle with the water founts and feeders in a circle around it. Put up your chick guard circle.

If you use a heat lamp, hang the bulb and set the feeders and waterers so that they will not be more than 24 inches from the center of the light from the lamp. Now, put up the guard. There is no set rule as to the size of the pen made by the chick guard. It should be large enough that the chicks

A comfortable home for baby chicks

can spread out if they wish and yet get back to the heat when they so desire. As the chicks grow, they will begin to regulate themselves according to the weather and the time of day, moving in when they get cool and out when they get too warm.

If you use a hover-type brooder, start it a day or so before the chicks arrive so you can make any necessary adjustments. Plug in the heat lamp to see that it lights properly. If it lights, unplug it until the morning of the day the chicks are expected.

Get a supply of feed. You will need about 10 pounds of chick starter for each chick up to 10 weeks of age. Feed bought in 100-pound bags is generally cheaper than feed purchased in smaller lots. But before you buy any feed, look at the tag on the sack and see what it contains. Some feeds contain chemicals such as arsenic, which is supposed to "sharpen appetites." There may be other additives equally unwholesome.

If so, consider making up your own chick starter ration. A good, natural chick feed is made of 2 parts cornmeal, 1 part each of wheat, cereal, oatmeal, and soy meal. This, with the addition of bonemeal and dried milk, will do quite well. As the chicks grow older we often substituted corn grits for the cornmeal. Green food, such as parsley or comfrey, is also excellent for chicks, as well as a few kale leaves for them to peck at. Let them have some fine grit, or sprinkle it on the feed.

WHEN YOUR CHICKS ARRIVE

When your chicks come from the hatchery, they are tender and easily chilled. Handle them with care, and put them under the brooder as soon as they arrive.

Put your starter rations in the feeders and water in the founts. Sprinkle some feed on cardboard or heavy paper so the chicks will see it. Sometimes it helps to pick up a few chicks and dip their beaks in the water and the feed to give them the idea. When some chicks eat and drink, the others soon learn from them.

Watch your chicks for a day or two. If they crowd under the heat, they are cold and won't go out to eat and drink. If they race to the edge of the guard, they are too hot. If they stay on one side of the source of heat, there is a draft. If they are scattered around the hover eating and drinking, the heat is just right.

Listen to your chicks. Peeping and complaining indicate that something is wrong.

Sanitation is important. Clean your founts daily with warm water and a sanitizing agent. When the feeders are empty, clean them before refilling. Watch both the feeders and waterers and remove any litter or debris that gets into them.

As your chicks get older, add feeders and waterers as needed. They should be up off the floor, with the lips of the troughs and founts just even with the top of the chick's back. (Place them on brick, or on platforms especially made for that purpose.)

AS YOUR CHICKS GROW

In warm weather, when the chicks get bigger, increase their area and use heat only at night and on rainy days. Once their dependency on the heat lamp is gone, they will simply ignore it and choose a sleeping corner of their own away from it.

By the time spring arrives, your young poultry will be ready to explore the chicken yard, looking for green supplements and insects. They should be able to handle regular grit. By the time they start laying, in summer, you should provide them with a box of crushed oyster shells so they will lay thick-shelled eggs. Chickens like kitchen tidbits and garden trimmings in season. Put them in a container outside so the poultry house stays free of odors and unwanted scraps.

How you feed your chickens will largely determine whether you have a profitable operation. By this I do not mean "commercial," but rather whether or not you will have a good supply of eggs and "eating" chickens sufficient to make your care and time worthwhile.

Commercial feed is expensive, but there are times when you will probably need to have it on hand to "fill in." The best way to feed chickens, of course, is to keep them as a part of a small farm operation, letting your flock have the run of a barnyard and fields. Under these circumstances, a small number can practically feed themselves during a major portion of the year.

Chickens need vast amounts of live protein daily. Free-ranging birds find this in the form of worms, flies, and beetles, collecting them to the benefit of the land they run on. At the same time, they enrich the land with their mineral-rich droppings. Chickens need to come in contact with the soil: It is in their nature to scratch in the earth for at least a part of their food. They also need the vital radiation of Earth itself, plenty of sunlight, and adequate shade to retire to during the hottest part of the day.

If your birds can't have access to fields, they should at least have a run, or better still, two runs that they can use alternately. Chickens will, in short order, trample your plot of land into a rather unlovely place.

If you have room for two runs, one can be used while the other "rests." While it is resting, grow it to a crop of quick-developing mustard, an excellent disinfectant when dug in green. Add some salt and soot to cleanse the soil even further.

Meanwhile, put your chickens into a scratch run. Litter this well with hay, leaves, or straw (even weeds will do if you have nothing else). About twice a day, sprinkle grain into the litter, a couple of handfuls per hen of oats, wheat, millet, buckwheat, or a mixture, along with herbs and garden vegetable discards. Scratching it into the litter provides healthful exercise, and such varied fare will give the birds important minerals.

Add to the litter from time to time, and change it once or twice a year. As the chickens scratch it about, they will break it up into valuable compost for the garden.

TO KEEP DOWN FLIES AND OTHER PESTS

At certain seasons, particularly in spring, when rains are frequent, flies can be very troublesome. Diatomaceous earth, an effective natural insecticide, can be fed to chickens as a means of fly control. Add a small amount to the feed of mature hens and you'll find that their droppings are less attractive to flies. Odors will noticeably diminish as well. For baby chicks, mix some in with the sand for their litter. Dust chicken roosts with diatomaceous earth to discourage mites.

For vermin on fowls and about hen roosts, steep pennyroyal in water to make an herb tea. Spray thoroughly over the roosts to cleanse them from lice. Pennyroyal is also effective against fleas.

Flies are particularly annoying when a change in the weather (from cloudy or unsettled to increasing moisture or rain) is imminent, stinging and swarming more than usual at this time. Chickens often indicate changing weather by rolling in dust or sand.

CHAMOMILE FOR CHICKS AND COMPOST

Dried chamomile, crumbled into the nesting hay for laying hens and scattered in the scratch and roosting area, will aid greatly in holding down odors. Litter containing chamomile, when added to the compost heap, speeds up the breakdown of other organic matter.

EGG PRODUCTION

A fancy poultry house is not necessary, but a clean one is essential for egg production. Keep sand on the floor with the addition of sawdust, preferably of pinewood, in wet and cold seasons. An occasional sprinkling with powdered lime will help to keep the place sweet and hold down odors. Keep feeders and waterers clean.

Do not overfeed your hens. Chickens are habitually sparse feeders, and overfat birds will not molt well, are often nervous, and may even become infertile.

Save yourself some money and also give your hens the benefit of "greenstuff" by arranging with a grocery store to save lettuce and cabbage trimmings, which your hens will eat avidly. While I advise rabbit producers not to feed such material because of sprays sometimes found on the outer layers of lettuce, I have never known such food to harm chickens.

Cut up discarded garden vegetables and feed them to your chickens, as well as any fallen fruit; the worm in it is often a special treat! In addition to table scraps, laying hens should always have access to oyster shells or a suitable substitute such as seashells, which you can pick up yourself and hammer into bits. Grains (corn, for example) may sometimes be gleaned from fields.

BUTCHERING THE COCKERELS

I know this isn't easy. By now those fluffy little chicks have come to seem like family. But killing them at the right time *is* necessary, and you must either do it yourself or have it done. Sometimes there is a friendly farmer in the vicinity who will do the deed for a modest fee.

After carefully scanning the flock and deciding which cockerels you want to keep, it is most practical to butcher the rest at about 2 months of age, when they are around 4½ pounds live weight. At this time the meat is very good and tender. Surplus hens you don't wish to keep can go just a little longer before the meat begins to toughen, but keeping chickens too long is a waste of feed.

After dressing, put the chickens in the freezer. The neck and giblets should be put into a small plastic food-storage bag and tucked inside the cavity. Wrap the chickens well to prevent freezer burn.

Space this over a couple of weekends so you can have fresh fried chicken every Sunday. It's almost a different bird from the store-bought article.

STARTING CHICKS FROM SCRATCH

With the cockerels neatly tucked into the freezer and the hens happily laying in the poultry house, we come at last to the topic we've been working toward from the very beginning: raising chicks naturally.

Realizing that incubators are, or, perhaps, must be the way for large commercial operations, I still insist that the best way to get the maximum health in chicks is to hatch them under the hen.

Eggshells are porous, which means the young and growing chick receives healthful radiations from its mother's body throughout the time of sitting. This leaves the eggs at hatching time lusty and strong of nerve. There is another factor to consider: The hen naturally has the gift of health selection and will push any unfit eggs out of her nesting box.

Of course you will need to keep a rooster. The breeding flock should have one for every 15 hens. Many times I have watched the barnyard drama play itself out. The rooster selects his own harem. He looks out for their interests, scratching diligently, if permitted, and calling to them when he finds a choice morsel. And they had better be his *own* hens that come at his call. The roosters divide up the territory, too, and squabbles sometimes result if one infringes on another's chosen area. In this small world, the "pecking order" is very real.

A rooster with his harem

Give the hens you have chosen a good diet at this time, being sure it is well balanced with lots of protein, greens, and minerals. They must have plenty of sunshine or cod-liver oil, necessary for strong eggshells and good hatchability.

HERB DELIGHTS

Two herbs eagerly sought by chickens are shepherd's purse, which possesses important astringent properties, and the common groundsel. Groundsel is rich in minerals, especially iron. Poultry will, if it's available, seek it out as a tonic.

THE BROODY HEN

As with most birds and animals, nature has decreed that early spring is the time for mating. Hens usually begin to show symptoms of broodiness around March or April. How can you tell a broody hen? Well, when she starts to stay on the laying nest for a couple of nights in a row, that's an indication. She may dart nervously about, ruffling her feathers and fussing peevishly.

Always "set" eggs (put them under a broody hen) on a date such that 21 days later they will be hatched when the moon is new and in a fruitful sign. The chicks will be hardier and grow faster than those born under an old moon, will come out of their shells close to the same time, and be strong and alert. They will mature rapidly and be good layers. *Llewellyn's Moon Sign Book and Gardening Almanac,* published annually, gives best dates for setting eggs.

The best signs are Cancer, Scorpio, and Pisces, the water signs. The second best signs are Taurus and Capricorn, the earth signs. If you plan for show stock, choose Libra, an air sign.

Chicks hatched under an old moon and in a barren sign will not show good

A broody hen with her suitor

results. They are apt to straggle out of their shells on different days and be weak and sluggish.

Chicks hatched in Gemini are not good layers. They will spend most of their time clucking and eating and will be active and restless, apt to fly over the fence and roam. Those hatched in Leo and Virgo will be of only average production. But chicks hatched in the water signs (the best of which is Cancer) will be maternal, quiet, and good layers. Save chicks hatched in this sign for your next generation of brood hens and you will find the desirable qualities considerably increased.

WHAT ABOUT NESTS?

When broodiness becomes evident, transfer the hen to a nest in the brooding room. Keep setting hens apart from the laying flock. This will keep other chickens from molesting them or adding new eggs to their nests.

Nests need not be elaborate. You can use apple boxes (with a few boards nailed across the top), nail kegs, or even sturdy cardboard boxes. Cut a round hole in one side about 4 inches from the floor to serve as a nest opening. Don't make it over-large, just big enough to be convenient for the hen to enter. There should be a few ventilation holes in the sides.

Remember, it is essential to provide moisture for the eggs in nests not located directly on the ground, so spread a bucketful of damp earth in each nest. Over the earth place a thin layer of straw or hay. Bank up both the dirt and the straw in the corners a bit, thus forming a depression in the center to keep the eggs together.

If parasites are a problem, drive them out of the nests of sitting hens by placing in the nest an egg that has been emptied and into which you have inserted a bit of sponge soaked in oil of eucalyptus.

BROODING

Years ago, before incubators for eggs came into general use, broodiness was a natural part of a hen's makeup. Nowadays, the modern hen may have to be encouraged a little the first time or so. At one time it was possible to obtain glass eggs, smooth and opaque and looking much like the real article. These were used to test a hen's intentions. Lacking these, probably your best bet is to use hard-boiled eggs. No, they won't hatch, but if she isn't serious they won't make too much of a mess, either.

It is best to move a broody biddy from the laying house to the brooding room at dusk. At this time, give her two or three of the hard-boiled eggs and shut her into the nest with a weighted box in front of the opening. Let her off of the nest for 15 or 20 minutes each day to eat, drink, exercise, and relieve herself. Then shut her in again. After a few days she should settle down.

In the meantime, while your broody biddy is being coaxed a bit to fulfill her role, you should be gathering and saving eggs. To keep them in good hatching condition for a week or more, place them where the temperature stays somewhere between 45 and 60 degrees. Place them in egg cartons, small end down. Do not wash them; this will remove the "protective bloom" that prevents the developing chick from drying out. The hen turns her eggs instinctively, so when preserving eggs for incubation you must turn them also.

Rotate the eggs daily. Remove the older ones to the refrigerator for use, and store the freshest just in case a hen or two decides to set.

When your broody hen has calmed down, remove the boiled eggs and give her a dozen fresh ones. This is your opportunity to choose a good day in a fruitful sign. At this point, most hens won't require any attention for the 21 days they will set. Just be sure that food, water, and a place for a quick dust bath are available each day when she comes off her nest.

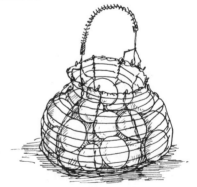

Remove older eggs to the refrigerator for use.

MONITORING AND CARE

Don't be disappointed if you have some failures, and do check on things rather often. You may have an occasional hen that lets her eggs get cold because she didn't return promptly, and this may cut down the hatching percentage. Sometimes an aggressive hen chases another hen off her nest and tries to take over. To avoid this, arrange the nests facing in different directions so that a hen will not become confused and get into another's nest by mistake. Hens aren't terribly bright!

Be sure there is plenty of space between the nests. This is important after the chicks hatch. A setting hen may peck to death a strange chick that climbs into her nest by mistake. You'll need room also because when the chicks hatch, you'll be putting a piece of board or some bedding in front of the nest opening to serve as a step for the babies. Disturb the hens as little as possible when the eggs begin to hatch.

Broody biddies that prove to have desirable maternal traits may be used for two or three years before you have to replace them. During this time, you will begin to know their individual characteristics. Remember, replace them in time with chicks born in the best signs and from the best mothers. The next generation will be greatly improved.

You may find, in time, that certain hens are quite amiable about adopting chicks from another hen, or even other fowl, should this prove necessary. A good way to persuade a reluctant hen to adopt new chicks is to smear the heads of all the chicks with vegetable oil. Now the hen can't tell the difference beween her own chicks and the new ones. Slip the new chicks under the hen at dusk.

Chicks don't always all hatch at the same time, but nature has provided for this. She stores enough food in the chick's body before it hatches so that the first-hatched chicks in a brood can wait while the late hatchers dry off and gain strength. When at last all are ready, mother hen will lead them proudly off the nest. Let food and drink be ready and waiting. At last we have come full circle.

Feed your new chicks as you did those in the brooder house when you first started your flock. Give them some skim milk, or a mixture of fish meal and water. And lucky the hen and her chicks that have the run of the outdoors and so round out their diet with greens and insects!

When the chicks are 5 or 6 weeks old and are well feathered out, remove them from their mothers, who are then returned to the laying flock. If the younger chickens cannot be kept separate, take care that the older ones do not keep them from getting enough feed.

PROBLEMS

✿ *Coccidiosis.* This ailment of the intestines is quite common in poultry, particularly when there is overcrowding, insufficient exercise, and lack of green food. Feed dandelion greens and young stinging nettles chopped together to prevent or cure coccidiosis in baby chicks. Give older chickens flaked garlic cloves, and continue this treatment for about 10 days. Exhausted chickens may be given drops of warm honey, which frequently acts as an immediate restorative.

✿ *Colds.* Give plenty of garlic internally. Withhold food for one day, then provide a laxative diet. When they are restored to health, remove poultry to fresh ground, and heavily lime the old run.

✿ *Worms.* A free-ranging farm hen should be free of worms, which are usually the result of artificial rearing, crowding, and inadequate diet. Symptoms vary with the type of worms: roundworms, tapeworms, hairworms, gapeworms, or pinworms. Give flaked cloves of garlic, or 2 or 3 drops of oil of eucalyptus in a little milk, to each hen. Give the remedy at the full moon.

✿ *General debility.* Most chickens that are allowed to range freely seldom show symptoms of illness. But when they do, there is an ancient remedy: the aloe plant, generally referred to as bitter aloes. In the Netherlands Antilles, where acres of the succulent plants are raised commercially, they are used to treat suspected illness or disease in chickens by throwing a live plant in the pond or watering tank from which the birds drink. The juice is also used for burns, and mixed with syrup to make cough medicine. Aloe plants may be purchased from nurseries in the United States.

✿ *Lice.* Mugwort is one of the most useful artemisias. Plant some in chicken yards to repel lice. Ingested, mugwort works as a vermifuge.

PISCES FOR DUCKS

The gestation period for duck eggs is 25 to 32 days. The usual number of ducklings is 9 to 12. I have always found the best breeding times for setting eggs to be Pisces, Cancer, and Scorpio, in that order. These are, of course, water signs. The earth signs of Taurus and Capricorn may also be used but are considered second best, particularly in the case of waterfowl such as ducks. Ducklings hatched in the sign of Pisces should be kept for future breeding stock.

All domestic ducks, except for the Muscovy, are descended from the wild mallard. There are no definite records telling us when ducks were first tamed but it must have been at least several thousand years ago.

Tame or domestic ducks come in three varieties: (1) those bred for meat, (2) those bred for eggs, and (3) those bred for their fancy appearance.

The most important meat ducks include the Pekin, Muscovy, black Cayuga, Rouen, and Aylesbury. We found the Muscovy breed especially to our liking because of its good eating and egg-laying qualities and because the females do not quack. Pekins are fine in every other respect, but they *are* noisy. If you live on a farm where they have ample space to roam and a pond to swim in, this is not too much of a problem, but under other conditions quacking is a definite nuisance.

Ducks are among the hardiest and easiest of all fowl to raise. They are a tremendous asset to place on a farm pond, as they help to cut down the algae (which do not aid fish growth, as plankton does).

It is not absolutely necessary to have a pond in order to keep ducks. For meat and egg purposes, a yard with a shallow cement or steel basin is adequate. It should be deep enough for the ducks to wade, wallow, and wash their feathers in, and should be continuously supplied with fresh, flowing water.

If you have sandy soil in your duck yard, so much the better; in any case, the soil must be well drained. Locate your yard on sloping land. If your space is small, don't plan on keeping too many ducks at a time. Ducks can be fairly sloppy and noisy, so don't crowd them.

Ducks are among the hardiest and easiest fowl to raise.

HOUSING

Housing for ducks is pretty much the same as it is for chickens, except that ducks don't use roosts. Keeping ducks and chickens together isn't very practical, though sometimes it can be done. Ducks, being more cold-hardy than chickens, do not need a building as tightly constructed. However, if they are housed in a building warmer than they need, they consume less food and develop less protective fat, two factors that can be an advantage to you. Simple nests, comfortably bedded and set 4 to 6 inches off the floor, are all they need.

HEN MOMS AND DUCK BABES

In many ways, it is easier to buy day-old ducklings than to try to raise them yourself, but you can have a breeding flock if you have the room and the inclination.

Wild ducks make excellent mothers, but most of the domestic breeds are temperamental. You may find that a broody chicken hen does the job much better. If your flock of ducks is small, gather as many fresh eggs as you can and store them in a cool, damp place (no longer than 5 days), until you have about 10. Turn them twice a day. At about dusk, slip a batch of duck eggs slowly under your broody hen. Be careful that she doesn't peck you; she'll think you're trying to steal *her* eggs even if she's sitting on nothing.

Most varieties of duck eggs hatch in about 28 days, except the Muscovy, which takes 34. Keep an eye on mother hen; she'll leave her nest at least once a day to eat and relieve herself, taking perhaps half an hour. This is your chance to sprinkle the duck eggs lightly with warm water, particularly necessary toward the end of the hatching period. Duck eggs need more moisture than chicken eggs do. Moist earth under the bedding, as for chicken eggs, is also helpful in maintaining humidity.

Be sure to turn the eggs once a day. Mother hen does this herself with her own eggs, but the duck eggs are a bit too big for her to handle.

Once the eggs are hatched, keep mother hen confined to a floorless cage about 3 by 3 feet in area. This should be raised high enough so the ducklings can crawl under the bottom to roam a little if they like. They won't go far. Chickens like to walk, but ducks don't care much for it. Confining the hen prevents her from exhausting the ducklings.

When the ducks are about 4 weeks old, release mother hen and let her roam at will with her adopted family. At 6 or 7 weeks, the ducklings are

A broody chicken hen may be a better mother to ducklings than their natural mother.

ready for their first swim. The hen will become violently upset when "her" ducks take to the water. She may need several days to calm down.

Feed young ducks much the same as you would young chicks, but they will grow far more rapidly, as I found to my sorrow when I tried to raise the two together under a brooder. The ducks stepped all over the chicks before I finally got the situation straightened out.

Ducks eat insects, snails, frogs, and fish. They also feed naturally on grains, grasses, and other forms of plant life. Free-ranging ducks find much of their own food, and a small flock will practically keep themselves. Free-ranging birds are also very healthy, particularly when they have access to water and such foods as cresses and water mint. They need more abundant green food than do chickens.

TAURUS FOR GEESE

Domestic geese are the descendants of the greylag goose of Europe. Growing much larger than their wild ancestors, domestics have almost lost their ability to fly. Now they can take only short hops into the air. Geese have long lives and in captivity occasionally reach more than 30 years of age.

Geese are highly intelligent. During the Vietnam War, the U.S. Army used geese to guard several key bridges in the Saigon area. They were considered more effective than watchdogs; the geese began to honk whether friend or foe approached, alerting soldier guards.

Goose eggs hatch in 27 to 33 days. If you want your geese to be good "watchdogs," plan for an average gestation period of 30 days. Set your eggs so that, by counting backward, the goslings will be hatched in Taurus, which is an earth sign ruling the throat.

The most popular breeds of geese are the Toulouse, Embden, African, Chinese, Pilgrim, Sebastopol, Canadian, and Egyptian. The best for farm use are generally considered to be the Toulouse or the Embden. If you want a real honker, choose the African: This breed is *very*

"Watch geese" are less likely than their four-legged counterparts to be swayed by treats.

noisy. The Toulouse is quite handsome, being gray with a white abdomen sweeping up to its tail. Geese are big birds; a mature Toulouse weighs between 20 and 25 pounds.

RAISING GEESE

Geese are even easier to keep than ducks, and during the spring and summer months you can permit them to subsist largely on pasturage. Geese keep weeds out of irrigation ditches, canals, and ponds, and control them in alfalfa and cotton fields and strawberry patches. They are very useful weeders for nurseries that grow trees, shrubs, and evergreens. Do not let them pasture in an orchard of young trees, however, for they will destroy the bark. (When the trees are older, they are fine.)

Geese are very hardy. A small house with an open entrance and a floor to keep out dampness is all they need. Insulation is necessary only in very cold areas. In southern states you may not need to provide them with shelter at all, but they must have shade, either natural or artificial. Geese are disease- and parasite-resistant. This bonus offsets somewhat their sloppiness, which is considerable when they are confined to a small area.

If you plan to allow your geese to range freely, figure on about 10 geese to an acre. Since they graze close to the ground, rotation will occasionally be necessary. In winter, increase their grain supplement to about 20 percent of their diet and plan on giving them some good legume hay or silage as well. Roughage may be supplied by vegetable tops and parings.

To fatten your goose for a holiday dinner, put it on a moist-mash fattening ration for a couple of months previous to slaughter. To make this feed extra fattening, mix equal portions of yellow cornmeal and oats with buttermilk or skim milk. Switch the goose to this diet gradually, letting it also have pasturage or other greens to prevent digestive problems. A period of a week or 10 days should be sufficient for the change.

MATING

Geese may mate for life; in this case, don't break up the happy couple. One gander, though, may cozy up to as many as four geese. If you want roast geese, bigamy is probably your best bet. Take special care during the mating season. Nasty fights can ensue, and segregation may be necessary to prevent bloodshed.

Be cautious, too. Ganders are naturally pugnacious and become more so at mating and rearing time. Be extra careful to keep children away from the geese.

THE HEN TO THE RESCUE

You may expect the female goose to lay an average of 25 eggs a year, over a period of about a month. This is too many for her to set; 15 is about all she can handle. Again, enter mother hen. These are *big* eggs, so don't give her more than 5. As with the duck eggs, they are too big for her to turn, so you will have to do it.

Let your hen start setting first, before the goose. Place her first 5 eggs under the hen and the second 5 under another hen. Let the goose set the rest. When the eggs under the goose are hatched, try slipping the hen-hatched goslings into her nest at night. Usually she will adopt them. If she doesn't, give them back to the hen. Keep a close watch on affairs at this point, making sure that all is well before you leave the goslings on their own.

When the eggs under your broody hen hatch, keep her confined as you did with the ducklings; goose eggs tend to hatch unevenly. If mother goose rejects them, keep the hen and goslings penned together for about 4 weeks. Up until they are 4 weeks old, take care that they do not get wet or even damp. Don't let them out on the grass unless you're sure it's completely dry. When they are about 1 month old and fully feathered, you can allow them to swim.

Feed for the goslings is much the same as feed for ducklings. Be sure they have plenty of fresh greens, plenty of fresh water, and provide food four times a day, as much as they will take.

MORE FROM YOUR GEESE

What about eating goose eggs? My advice: Hatch them or sell them to another goose fancier but don't bother to eat them unless you're *really* hungry. I have used them to cook with, however, and in a pinch they do reasonably well.

What about goose feathers? Ah, that's something else. Once upon a time, no bride in a country district ever went to her new home without a pair of goose-down pillows. My mother gave me a pair, and I still think it's a wonderful custom. Geese should be plucked in the increase of the moon.

PIGEONS

Pigeons have been one of man's closest associates for more than 4,000 years. They have been useful for laboratory and research purposes. Some varieties are selected for their fancy colors and interesting forms, and many of these are truly gorgeous. The sign of Libra enhances this beauty. Other types of pigeons have been selected and bred for their endurance and homing instinct. Still others are bred for food purposes; the young pigeons, or squabs, are especially delectable. Taurus is the best sign for this.

Pigeons readily adapt to living under a variety of conditions. Their diet is simple. Pigeons are easily tamed and are normally free of objectionable odors. The noises they make are not loud or harsh. Because of these factors they can be kept in urban locations. They require little space and you can obtain them at reasonable cost. Buy pigeons when the moon is in good aspect to Venus.

FEEDING PIGEONS

A mixture of whole grains, some good hard grit, and plenty of fresh, clean water are all they need. Place the feed, grit, and water in containers spaced some distance apart. Containers should be so made that very little, if any, dirt or filth can get into the feed and water.

You can get pellets (checkers) for pigeons. These are a mixture of finely ground grains and other food particles. However, if only pellets are fed, they may produce loose droppings. A mixture of grain and pellets corrects this condition. You may feed at a 50-50 ratio or as low as 25 percent grain and 75 percent pellets.

Pellets are desirable for breeders feeding squabs in the nest because they are rapidly digested, enabling the parent to regurgitate a considerable quantity of partially digested feed (called pigeon milk) in a given day. It's a good idea to feed pellets in the morning and grain in the evening, thus fortifying the bird during the time when it is on the nest or roost at night.

Place each feed in a separate container, cafeteria-style. This allows the bird to choose and satisfy its own desires. When you present feed like this, use feeders that prevent waste. Birds given free choice have a tendency to be rather sloppy as they search for the ideal kernel or food particle. The best practice is to keep some feed before your birds at all times. Mature birds producing young for slaughter should *never* be without feed.

Mixed grains for pigeons can be field or Canadian peas, together with flint corn, whole wheat, and the various grain sorghums. You can get mixed pigeon grains from most feed companies. A good mixture is free of dust and has no broken or cracked kernels. Pigeons prefer the whole grains and will reject cracked ground corn.

Protect feeders or other receptacles from the weather. The feeders should be placed inside the coop or house rather than out in the fly pen. Remove any grain that is accidentally moistened, since moldy grain is a source of much trouble. Protect containers for storing feed and keep them tightly closed. They should be made of metal so that rats or mice cannot gain entry.

Grit is to pigeons what teeth are to people. They need a supply of small gravel or granite grit at all times. They will not consume very much, but it should be available. Also provide a small amount of oyster shell or similar material for good eggshell formation. Oyster shell and grit should not be confused. They are both needed, but for different purposes. A balanced mineral and grit mixture is essential to avoid sterile eggs, poor hatchability, crippled squabs, and a host of other problems. You can purchase some from feed stores or order through the mail.

Be sure there is plenty of clean, fresh water at all times. Pigeons drink by swallowing (as a person does) and not by dipping their beak (as chickens do). Thus, make sure that water is at least ¾ of an inch deep at all times. Don't put the water in open pans or the pigeons will bathe in them. Wash and disinfect waterers at least once a week to prevent disease.

HOUSING YOUR PIGEONS

Breeding pigeons need a sound, leakproof house, but it need not be fancy. A corner of the barn, an unused garage, or a chicken house will do. Each pair of pigeons should have about 4 square feet of space. The pigeon house should have either a cement or a solid wood floor, never earth. It must also be dry and free of drafts. A glass window that can be opened and shut will serve for light and entrance.

If you live in a rural area, you may be able to let your pigeons fly free, but I don't advise doing this otherwise. Pigeons can be a great nuisance, especially to gardening neighbors. Build a flying pen in front of your breeding house. A pen about 10 feet wide, 7 or 8 feet tall, and about 15 feet long will be adequate for a flock of 12 to 15 pairs. To keep out grain-eating sparrows, use 1-inch-mesh wire.

Build several narrow plank walkways around the pen near the top so your pigeons can strut and sun. When the weather is fine, place a flat bath pan in the flying pen and let the birds splash about several times weekly.

BREEDING LOFTS

In a warm climate, you can use open-front breeding lofts. In colder areas, close up the fronts and also shut doors and windows to keep out the cold. Given this care, the pigeons will go right on breeding.

Don't overcrowd your breeding loft. A loft 10 to 12 feet square houses 12 to 15 pairs in comfort. The loft should be rat- and mouseproof with fine wire. Try to have the door (with the upper portion glass if possible), window, and breeding loft facing south to get full sunshine. You can then set up your nest boxes for breeding.

PIGEONS MATE IN PAIRS

Pigeons mate in pairs and generally remain "wedded" for life. After a pair have gone through the courting stage and have mated, they are ready to build a nest and hatch their young. They remain true to each other as long as they live or as long as they are allowed to remain together.

A word of caution: Some dealers are rather cautious in talking about mated pairs; they speak of selling pairs of pigeons without saying anything about mating. *Insist on getting mated pairs when you buy breeders.*

Birds that have mated can be shipped long distances without breaking the matings; occasionally, though, a pair break their mating from relocation or the influence of unmated outsiders.

You can locate breeding stock through pigeon clubs, by asking your grain or feed dealer, or by contacting your county agent. Buying directly from a breeder is always best.

BREEDING HABITS

Pigeons have unique breeding habits. When a male has selected the female he desires for his mate, he struts around his favorite, coos to her, and evidently tries to show her what a grand fellow he is. The female, if attracted, becomes friendly and the two "bill" each other as if they were exchanging kisses.

Pigeons in love

Things then get cozy. The two select a spot and build their nest. The cock bird becomes anxious for the hen to begin laying. If she doesn't take up her duties promptly, he drives her into the nest. He will even "talk" angrily to her and strike her with his wings. Finally the hen goes to the nest and lays an egg. She skips a day and then lays a second egg, and that's that for now.

As soon as the first egg is laid, incubation begins. The hen occupies the nest from about 4 o'clock in the afternoon until 10 the next morning. The cock then obligingly sits while his mate eats and rests. In this manner the incubation goes on, and at the end of about 17 days the egg that was laid first hatches. About 24 hours later, the second egg hatches. The incubation period ranges from 16 to 20 days.

One of the young birds will be a full day older than the other. Almost always the first hatched is a male; the second is usually a female.

The adult birds begin immediately to feed their young, which make rapid growth. The young, known as squabs, are kept well filled by their parents with a fluid called pigeon milk. This is a combination of secretions from the parents' crops and partially digested feed. On this diet the young seem to grow before your eyes.

In a few days, the hen is ready to lay again. Because of this, each breeding pair should have two nest boxes. During a large part of the year they will be sitting on a pair of eggs in one nest and have a pair of large squabs in the other.

Feeding a squab

Orange or egg crates make inexpensive nest boxes. Stack them on their sides and slip a plywood or masonite bottom in each nest box to cover cracks.

Provide a loose rack for nesting materials in the loft. This can be hay, straw, pine needles, grass, or tobacco stems. You may even build up some of the nests for the birds. Cheap, pulp paper nest bowls are available, one for each nest, at feed stores. Use them for several months; then discard and replace. Squabbing pigeons like their breeding loft slightly darkened.

SQUABS

When squabs are 28 days old and well feathered under the wings, they are ready to be taken and dressed for food. They are heavier at this age than they will ever be again in their lives. If they are not taken from the nest at about this time, the old birds, desiring to start another pair of eggs, push the squabs out. They fall on the floor of the loft so fat they can hardly get about. Here they become lean while learning to eat for themselves and soon are sleek and trim (but less tender and tasty).

Squabs should always be dry-picked, never scalded. After picking, wash the squabs and leave them to cool in a pail of cold water for about half an hour. For home use, cut off the feet and head and eviscerate, leaving the carcass whole, though sometimes you can slit it down the back to spread. Those not intended for immediate 'use should be bagged or wrapped in pairs and placed in the freezer. Frozen squabs, if well wrapped, will keep well for up to 9 months.

PIGEON PROBLEMS

✿ ***Canker (trichomoniasis).*** This is a killer of young squabs at about 2 weeks of age. It shows up as a yellowish white pustule or crust in the throat of the young bird. This eventually grows large enough to close off the windpipe and produce suffocation.

The treatment most widely used is the application of a mild astringent to the area of the bird's throat as soon as canker appears. Check with your county extension agent if you notice signs of canker, and use the preparation he or she suggests.

✿ ***Mites and lice.*** Lime the pens and fumigate yearly with cayenne pepper to hold down vermin. Young squabs should have a little oil of eucalyptus applied around the neck, under the beak, and rubbed into the top of the head. (Avoid the eyes.) Examine the birds 7 to 10 days later and repeat if the first treatment was not 100 percent effective.

Insecticidal sprays can be used for the large loft (ask your county agent or veterinarian for recommended products). In this case, the equipment as well as the birds should be sprayed.

If disease appears in your flock, the first step is to get a reliable diagnosis from your state animal health laboratory or a local veterinarian specializing in small animals and pets.

A mixture of basalt and cumin is a good tonic to give pigeons occasionally. They also like caraway seed, which, for pigeons that fly free, will lure them back to the loft.

> **HELPFUL SUGGESTIONS**
>
> Keep gravel on the floor of the fly pen or on the ground where fly pens are located to keep the surface clean and presentable. Rain will wash the droppings through coarse gravel and into the soil. Elevated fly pens with wire floors are also desirable.

PIGEON PESTS

Do not let cats gain access to the nesting area; they will kill birds and rob nests. If coarse-mesh wire is used for the fly pens (1-inch mesh or larger), use window screening or nail screening to a height of 18 inches up from the ground to prevent cats from reaching through and catching birds walking near the fly pen walls.

Prevent the entry of rats, mice, and possibly snakes by having floors of tin or other metal, and small-mesh wire (not larger than ¼ inch) over all openings in the building, including side walls and the top of the fly pen.

To prevent youngsters from prowling and leaving doors open, equip your coop with locks and keep it locked at all times. (This also protects against thieves.)

Keep your pigeon coop clean at all times. An unsightly or neglected coop can become an objectionable neighborhood nuisance.

Plant tall-growing flowers such as sunflowers and hollyhocks around the pigeon coop in order to afford some screening from prying eyes. In time, trees or shrubbery can be planted instead of the annuals.

CHAPTER 5

THE RABBIT, SYMBOL OF THE MOON

Ancient peoples considered the rabbit a symbol of the moon.
In Egypt it was the sign of birth and new life.

When or why the rabbit first was considered symbolic of the moon is not definitely known. Perhaps the rabbit (or hare) later became an Easter symbol because the moon determines the date of Easter. The rabbit is ruled (that is, influenced) by Mercury, Venus, the moon, and Jupiter, in that order. Buy or sell rabbits when the moon is not aspected with Venus.

Rabbits require but little space and are very easy to keep. They neither crow nor bark. Little capital is needed to get started, and they are efficient meat producers. Rabbits grow out rapidly, and butchering a rabbit is similar to butchering a chicken.

Rabbit raising is especially adapted to small farms and rural areas where other livestock projects may not be practical. Rabbits are mostly raised for meat, but the skins do have some commercial value.

WHICH BREED?

Decide whether you want to raise rabbits for meat, fur, or show, and select the breed best suited to this choice. Mature animals of the small breeds weigh 3 to 4 pounds; medium breeds, 9 to 12 pounds; and large breeds, 14 to 16 pounds. The American Rabbit Breeders Association can tell you the standards for all breeds and has information about their numerous varieties.

CHOOSING FOUNDATION STOCK

If you're planning to raise rabbits primarily for meat, it isn't necessary to have purebred registered stock, but select *good* stock or you'll be wasting your time and effort. The essentials are health, vigor, longevity, ability to reproduce, and desirable type and conformation.

Begin on a small scale, expanding as you gain experience. One buck and three does will supply enough rabbit fryers for an average family. Select young animals, so you can become better acquainted with your rabbits before they reach production age.

National, state, and local rabbit breeders organizations can furnish names and addresses of breeders. Try to deal directly with reliable breeders who stand behind the stock they offer.

YOUR HOUSING AND EQUIPMENT

The building and equipment needs for a rabbitry depend on local building regulations, climate conditions, the size of the operation, and the amount of money you can invest. Is there a successful rabbit breeder in your area? Ask your county agent for leads.

Mature rabbits need individual hutches, about 2 feet high and no more than 2½ feet deep. Make the hutches 3 feet long for small breeds, 4 feet for medium breeds, and 6 feet for large breeds. (These figures are for inside measurements.) Several types of metal and all-wire hutches are available on the market, or you can construct hutches of scrap lumber and hardware cloth or wire mesh. Plans and specifications for them can be obtained through advertisements in various rabbit journals.

Wire-mesh flooring is used extensively in commercial rabbitries, where self-cleaning hutches are desirable. If you use it, examine the surface for sharp points. Always put the smooth surface on top. Solid and slat flooring, or a combination of solid flooring at the front and a strip of wire mesh or slats at the back, are also good choices in hutch construction. For slat flooring, use 1-inch hardwood slats and space them ⅝ or ¾ inch apart. Solid floors should slope slightly from the front or the rear of the hutch to provide drainage.

In mild climates, place hutches outdoors in the shade of trees or buildings, or under a lath superstructure. Sunlight in the rabbitry will help maintain sanitation, but your rabbits should have a choice between shade and sunlight.

During very hot weather it may be necessary to provide some additional cooling measures such as overhead sprinklers, or foggers placed within the building. Be sure your building is adequately ventilated and that the rabbits receive the benefit of prevailing breezes.

In colder climates, the hutches may be kept outside for perhaps six months of the year. In cold weather, put them in a corner of the barn or shed, or protect them by placing them in buildings that open to the south or east. During stormy weather, close these buildings with curtains or panels.

UNDER THE HUTCHES

Expect a 10- to 12-pound doe and her young to produce about 6 cubic feet of manure a year. This is the best worm feed there is. Furthermore, it's a simple matter to raise earthworms directly under the cages.

Save and compost any manure not used for this purpose. An average sample is rich enough in nitrogen to produce good heating in the compost heap. It is one of the finest soil builders for the garden, but don't use large applications of it fresh.

Sawdust under rabbit cages soaks up urine and cuts down on the smell of ammonia. The sawdust, eventually tilled into the garden, will keep the soil there loose and workable.

NEST BOXES

A good nest box provides seclusion for the doe when she gives birth and protection for the litter. It is simple to clean and maintain, a comfortable size, and provides adequate drainage and ventilation. It should be accessible to the young when they are large enough to leave and return to the nest.

Make an inexpensive nest box from a sturdy packing box (an apple box is ideal). Cut an opening in one end of the box, or remove a portion of one end, to provide easier access for the doe and her young. In a warm climate, you can remove one end entirely and replace it with removable slats. Leave the slats in place until the young rabbits are big enough to come and go.

Nail kegs also make good nest boxes. Nail a board across the open end of the keg, covering one-third to one-half of the opening. To keep the keg from rolling, extend the board a few inches beyond the edge of the opening. Be sure to drill several 1-inch holes in the closed end of the keg for ventilation.

During cold weather, young rabbits may need more protection than the standard nest box offers. A standard box lined with two or three layers of corrugated cardboard makes a winter nest box. Drill two or three holes in the lid for ventilation, then fill the box with clean straw so the doe can burrow a cavity for a nest.

FEEDING EQUIPMENT

Use feed crocks, troughs, hoppers, and hay mangers that are large enough to hold several feedings. You want a type that prevents waste and contamination of feed.

Feed pellets or whole grain from crocks especially designed for rabbits. These crocks, usually made of heavy earthenware, are not easily tipped and have a lip that prevents the rabbits from scratching out feed and wasting it.

When hay or green feed is included in the ration, incorporate hay mangers into the hutches. Save space by having one manger serve two hutches. Hay waste can be reduced by placing troughs under the mangers. (You can put supplemental grains in these troughs, too.)

Hoppers save considerable time and labor when they are designed for self-feeding. Use them for feeding pregnant does, does with suckling litters, and rabbits being conditioned for slaughter. Make hoppers from cans, cardboard, or wood. You can make a good, inexpensive feeder from a 5-gallon can.

WATERING EQUIPMENT

Rabbits must have clean, fresh water at all times. Coccidiosis (a fungal disease) is in large part traced to fouled water. An automatic watering system will eliminate much of the work and reduces coccidiosis and other problems. If you don't want the expense, crocks or coffee cans will work in small rabbitries. Coffee cans are especially useful during cold weather because you can easily break up any ice and remove it.

FEEDS

Feed is probably the largest item of expense in raising rabbits. We always used a commercial brand, a "bunny" type for the youngsters, rabbit food for the adults. We supplemented this with such grains or garden produce as we could raise ourselves.

HAY

Hay supplies the bulk, or fiber, required for a balanced rabbit ration. High-quality hay also provides much of the protein rabbits need.

Legume hays (alfalfa, clover, lespedeza, cowpea, vetch, and peanut) are all palatable and high in protein. Grass hays, including timothy, prairie grass, johnsongrass, Sudan grass, and carpet grass, are less palatable and contain about half as much protein, but they are valuable for feeding when legumes are not available. When grass hays are fed, include additional protein supplement in the ration.

Fine-stemmed, leafy, well-cured hay can be fed whole. Chop coarse hay into 3- to 4-inch lengths. This will reduce waste and make the hay more convenient for feeding.

Caution: Never, *never* feed spoiled hay. In all likelihood it will kill your rabbits. Be especially careful of sweet clover hay. It is toxic to animals when spoiled, causing both external and internal bleeding.

GRAIN

If you have the room, there are a number of grains you can grow yourself: oats, wheat, barley, sorghum, buckwheat, rye, and soft varieties of corn. Feed any of these whole or milled. To prevent waste, feed flinty varieties of corn in meal form. Grains are similar in food value; you can substitute one for another without greatly altering the nutritive value of the ration.

By-products from grain manufacturing, bran, middlings, and shorts, for example, can be included in meal mixtures and pellets.

SUPPLEMENTS

Soybean, peanut, sesame, and linseed meals are rich in protein, and desirable for balancing pelleted rations. Protein meal should not be mixed with whole grains, because much of the meal will settle out and go to waste. If whole grains are fed, supply the protein in flake, cake, or pelleted form.

Rabbits also require salt. Put small salt blocks or spools in the hutches so the rabbits can feed at will. Or you can add 0.5 to 1 percent of salt to mixed feeds or pellets. In areas where the soil is deficient in mineral elements, give the rabbits the same type of mineralized salt that you feed to other farm animals.

PELLETED FEED

There are several brands of pelleted feed. Pellets require little storage space and are easily fed. In some areas, they may be the only rabbit feed available. There are two general types: all-grain pellets to be fed with hay, and complete pellets (green pellets) which usually contain all the elements necessary for a balanced ration. Follow the advice of the manufacturer when you feed pellets.

Pellets should be no larger than 3⁄16 inch in diameter and ¼ inch long. If they are larger, small rabbits will bite off part of a pellet and waste the rest.

MISCELLANEOUS FEEDS

Rapid-growing plants, grasses, palatable weeds, cereal grains, and leafy garden vegetables free of insecticides are all high in vitamins, minerals, and proteins. These plants make excellent feed for your breeding herd. *Never* feed your rabbits lettuce or cabbage trimmings from a grocery store. The outer leaves are apt to have residues of poisonous sprays, which will kill your rabbits.

> **CAUTION**
>
> When feeding weeds or wild plants, obtain a list of the poisonous plants found in your area from your police department, state experiment station, county agent, or state university, and learn to recognize them.

Dandelion, a weed of open places, is blood cleansing and tonic. Wild rabbits seek out the leaves of thistle and brambles for their medicinal value.

Root crops, such as carrots, turnips, and beets, are desirable rabbit feed throughout the year and are particularly fine to offer in winter, when green feed is not available.

Fresh green feeds and root crops should be considered as supplements to the regular diet. Use them in the ration when they fit into your management program. Green feeds and root crops should be fed sparingly to rabbits that are unaccustomed to them.

Dry bread and other table waste (except meat and greasy or sour foods) are acceptable to most rabbits. These foods add variety to a diet when they supplement the regular ration.

We believe that rabbits, like all other animals, need herbs. Dried stinging nettle is helpful in maintaining health for rabbits of all ages. Lemon balm helps the flow of milk in the nursing doe.

Aromatic herbs such as thyme, sage, lavender, and rosemary are good, as is the sharp, medicinal mustard. Add dried chamomile to the hay in the next boxes to keep down odors.

Rabbits eating stinging nettle

RE-INGESTING

Rabbits re-ingest part of their food, usually in the early morning when they are unobserved. They re-ingest only the soft material that has passed through the digestive tract. This trait is normal in rabbits and may even enhance the nutritive value of the food by virtue of the second chance at digestion.

FEED APPROPRIATELY

Select rations that are suitable for the needs of your rabbits. Dry does, herd bucks, and junior does and bucks need rations that will keep them in good breeding condition. Just because they are not "producing" doesn't mean that you can skimp on feed. However, pregnant does and does with litters do need a higher proportion of concentrate in their rations.

FEEDING FOR MAINTENANCE

Junior does and bucks, mature dry does, and herd bucks not in service but in good physical condition can be maintained on hay alone if fine-stemmed, leafy legume hay is fed. When grass hay or a coarse legume is fed, give each 8-pound animal 2 ounces (¼ cup) of all-grain pellets or grain-protein mixture several times a week. Adjust the amount of concentrate in the ration for rabbits of other weights.

Feed herd bucks in service the same amount of concentrate and give them free access to choice hay.

FEEDING PREGNANT DOES AND DOES WITH LITTERS

To feed a doe properly, you must know definitely whether she has conceived. Palpating 14 days after breeding is a quick and accurate method of determining this. (See "Determining Pregnancy" on page 80.)

After the mating, keep the doe in breeding condition on good-quality hay. If the doe has failed to conceive, as determined by palpation 14 days after breeding, breed her again and keep her on a maintenance ration until she is diagnosed as pregnant. At this time, give her all the concentrates she can eat in addition to good-quality hay. All-grain pellets or a grain-protein mixture can be fed with the hay, or you can feed a complete pelleted ration and no hay. Feeding medicinal herbs at this time is also beneficial to help her maintain health and vigor. They will also have a good effect on the young she is carrying.

Be sure to change over to the new ration gradually. Sudden changes in rations during the gestation period may cause some does to go "off feed."

After the doe kindles, feed her as you did during pregnancy. Keep her on the high-concentrate ration until the young are weaned. Give additional feed as the litter develops.

BREEDING

The gestation period of rabbits varies slightly. Some litters may be kindled as early as the 29th day or as late as the 35th day, but 98 percent of the normal litters are kindled between the 30th and 33rd day. Figure on an average of 30 days, and arrange the mating on a day that will bring the litter when the moon is in the first quarter, preferably in Cancer, Scorpio, Pisces, Taurus, or Capricorn, in that order.

The right age of bucks and does for the first mating depends on the breed and individual development. Small breeds are sexually mature at a younger age than are medium and large breeds. Individual rabbits within a breed will develop more rapidly than others. We like to breed the does of the Californian or New Zealand breeds at 5 months and the bucks at 6 months. Mate does when they reach maturity; if mating is delayed too long, breeding difficulties may arise.

Does maintained in good physical condition will provide excellent litters until they are 2½ to 3 years old. If your principal interest is meat and fur production, you can work your breeding animals throughout the year. With a gestation period of 30 to 32 days, and a nursing period of 8 weeks, a doe can produce four litters in a 12-month period.

MATING

Restlessness and nervousness are signs that the doe is ready for mating. She will rub her chin on feeding and watering equipment and attempt to join rabbits in nearby hutches.

Always take the doe to the buck's hutch for service. Mating should occur almost immediately when you place the doe in with the buck. When the mating is completed, return the doe to her own hutch. Make a record of the date of mating and the name or number of the buck and doe. Try to maintain one buck for each 10 breeding does. Mature, vigorous bucks can be used several times a day for a short period.

DETERMINING PREGNANCY

It's important to use an accurate method of determining pregnancy. Test mating (that is, placing the doe in the buck's hutch periodically) is not accurate because some does will accept service when pregnant and others may refuse when not pregnant.

Pregnancy can be quickly and accurately determined by palpating the doe 12 to 14 days after mating.

Hold the ears and the fold of skin over the shoulders in the right hand; place the left hand between the hind legs, slightly in front of the pelvis. Placing the thumb on the right side and the fingers on the left side of the abdomen, exert light pressure, and move the fingers and thumb gently back and forth. Be careful to handle the doe gently, using only light pressure on the abdominal cavity.

When the doe is pregnant, you should be able to distinguish the embryos as marble-shaped forms as they slip between the thumb and fingers. Does diagnosed as pregnant should be put on a pregnancy ration. Rebreed nonpregnant does and keep them on a maintenance ration until they are pregnant.

Until you gain experience, repalpate does diagnosed as nonpregnant a week later. If the first palpation was incorrect, the doe can still be put on a pregnancy ration and provided with a nest box at the proper kindling time. As with most other endeavors, your skill in diagnosis will improve with practice.

POINTERS ON BREEDING

✿ Always take the doe to the buck's hutch, never the other way around, or a fight may ensue.

✿ Does that produce fewer than seven healthy babies, two litters in a row, should be replaced.

✿ Replace the buck if his record reveals low production, or his offspring show poor feed conversion or a poor rate of gain.

✿ Through correct breeding time and careful selection, save replacement stock for expansion as needed. Keep your cages filled with working does and active bucks. The herd should be constantly improved by culling low producers.

> **BREEDING TIP**
>
> If you have a doe that is slow to breed, try this: Remove all feed from the doe for about three days before breeding and give her nothing but carrots and plenty of fresh drinking water. You should see results.

✿ The average production life of a good doe is 2½ to 3 years. One young replacement doe should be saved each month for each 12 working does.

✿ A good buck may be used for 1 to 2 years. Save young bucks for replacement.

✿ Purchase good bucks occasionally from other herds to prevent inbreeding.

CARING FOR THE LITTER

Important: Place a nest box in the hutch 27 days after the doe is mated. For a day or two before kindling, the doe starts to pull fur from her body to line the nest. Usually she will consume less feed than normal. Give her small quantities of green feed to benefit her digestive system. Lemon balm is good for her to nibble on.

Most litters are kindled at night. Complications are rare when the doe is in good condition. After kindling, the doe may be restless. Do not disturb her until she has quieted down.

Sometimes does fail to pull fur to cover their litters, or they kindle their litters on the hutch floor. Should this occur, arrange the bedding material to make a comfortable nest and pull (gently) enough fur from the doe's body to cover the litter. You may be able to save the young rabbits by warming them if they have been chilled, even if they appear lifeless. I learned always to keep some extra fur on hand in case this happened.

Inspect the litter the day after kindling. Remove any deformed, undersized, or dead young from the nest box. I was always very careful and quiet when doing this and found the does did not object. As time went on and we raised our own breeding does, this and any other handling necessary became even easier. Our animals, accustomed to our touch almost from birth, had no fear of us at all.

We even found it possible occasionally to transfer baby rabbits from a large litter to a foster mother with a small litter. For meat and fur production, a litter of 7 to 9 is desirable. Adjusting the number of young to the capacity of the doe promotes more uniform development. However, for best results, the young rabbits should not vary more than 3 or 4 days in age when they are transferred.

Young rabbits open their eyes 10 or 11 days after birth. The eyes of baby rabbits occasionally become infected and fail to open normally. Treated promptly, the young rabbits usually recover without any permanent eye injury.

To treat the infection (and this is good for other young animals as well), bathe the inflamed and encrusted eyelids with warm water. I always keep dried borage on hand and steep it in the warm water, straining it out before using. When the tissue softens, separate the lids with a slight pressure. If pus is present on succeeding days, repeat the treatment.

WEANING

Your rabbits start coming out of the nest to eat feed when they are 19 or 20 days old. Young that come out of the nest sooner may not be getting enough milk, or the nest may be too warm.

The doe usually nurses her young at night or in the early evening and morning hours. Should the litter become divided, the doe is likely to nurse either the young in the nest or those on the hutch floor. She will not nurse both groups, nor will she pick up the young and return them to the nest. You must keep an eye on things.

Leave the young rabbits with the doe until they are 8 weeks old. By that time the milk supply has begun to decrease and the young will be accustomed to eating other feed. Fryer rabbits should be in marketing condition by the time they are weaned at 8 weeks.

Young rabbits are not born "tame." When they first come out of the nest box they will be just as much afraid of humans as are their wild prototypes. You must work with them gradually to overcome this fear, and considerable time and patience are required. This doesn't matter greatly if they're intended for slaughter, but if you wish to keep them for breeding animals, it's an advantage to overcome their nervousness from the very beginning.

RECORDS

Even if you are keeping only a few does and a buck to provide meat for the family larder, it is a good idea to keep records. This will enable you to keep track of breeding, kindling, and weaning operations, to cull unproductive animals, and select desirable breeding stock. Records cards can usually be obtained from feed mills or firms dealing in rabbitry supplies.

HOT WEATHER CARE

Newborn litters and does in advanced pregnancy suffer most from high temperatures. Heat discomfort in young animals causes extreme restlessness. In older animals, heat suffering is indicated by rapid respiration, excessive moisture around the mouth, and even occasional hemorrhaging from the nostrils.

Move rabbits that show symptoms of heat suffering to a quiet, well-ventilated place. Give them a feed sack moistened with cold water to lie on. Place in the hutches water crocks or large bottles filled with ice to help keep the rabbits cool.

During hot weather it may be difficult to keep the young rabbits comfortable in the nest box. A cooling basket of small-wire mesh will provide some relief. Cooling baskets are useful from the time the young are kindled until they are large enough to get out of the nest by themselves.

During the hot part of the day, place the young in the basket and hang it near the top of the hutch out of direct sunlight. In the evening, return the litter to the nest box. If high temperatures continue throughout the night, return the young to the cooling basket after they have nursed. Allow them to nurse again in the morning.

SLAUGHTERING

Rabbit meat is *good* meat, white, tender, easily digested, high in protein, and delicious. For many people, this is the primary reason for raising rabbits. The rabbits you raise yourself will be *undrugged,* healthy animals fed on wholesome grains, hays, and herbs. You will know exactly what you are eating. Tularemia is not a risk in domestic rabbit meat because the rabbits never touch the ground. This danger is almost always present, though, from eating wild rabbit.

Always slaughter in clean, sanitary quarters. The most humane method I know of is first to render the rabbit unconscious by dislocating the neck or by stunning with a sharp blow to the base of the skull.

To dislocate the neck, hold the animal by its hind legs with your left hand. Place the thumb of your right hand on the neck just behind the ears and place your fingers under the chin. Stretch the animal by pushing down on the neck with your right hand. Press down with the thumb, then raise the animal's head with a quick movement to dislocate the neck. This method is instantaneous and painless when done correctly.

Suspend the carcass by inserting a hook between the tendon and bone of the right leg. Insert the hook just above the hock. Remove the head immediately to permit thorough bleeding. Cut off the tail, the front feet, and the free rear leg at the hock joint. Cut into the skin just below the hock of the suspended leg, then slit open the skin on the inside of the leg to the base of the tail. Continue the incision to the hock of the left leg. Separate the edges of the skin from the carcass and pull the skin down over the animal. Leave as much fat on the carcass as possible.

After skinning, make a slit along the median line of the belly. Remove the entrails and gallbladder, but leave the liver and kidneys in place. Unhook the suspended carcass and remove the right hind leg at the hock.

Wash the carcass in cold water. Brush the neck thoroughly in water to remove the blood. Do not leave the carcass in water more than 15 minutes. (The carcass will absorb water if soaked for a prolonged period.) Chill the carcass in a refrigerated cooler.

Use a sharp knife to cut up the carcass. Do not use a cleaver; it may splinter the bones.

Slaughter rabbits in the same signs as you would for goat kids: the first three days of the full moon, and never in the sign of Leo.

SKINS

To save the skins, shape them while they are still warm. Place the skins, flesh side out, on wire or board shapers (stretchers). Place the forepart of the skin over the narrow end of the shaper and make sure the legs are all on the same side. Remove all wrinkles from the skin.

The day after skinning, examine the pelts. Make sure that the edges are dry and flat, that the skin of the front feet is straightened out, and that all patches of fat have been removed.

Skins must be thoroughly dried before you pack them, but do not dry the skins in the sun or by artificial heat. Hang them up so air can circulate freely. During warm weather, sprinkle the skins with naphtha flakes. Do not use salt to cure rabbit skins.

CONTROLLING DISEASES

Sanitation in the rabbitry is the best method of disease control. Remove manure, soiled bedding, and contaminated feed daily. Wash the watering and feeding equipment frequently in hot soapy water. Rinse in clear water, drain well, and place in full sunlight to dry.

Isolate any animals suspected of being diseased. Leave them in isolation for at least 2 weeks, or until you can determine whether they are dangerous to the health of the herd. Newly acquired rabbits, and those returned from shows, should be placed in quarantine for 2 weeks. Bury or burn dead animals.

Many rabbit diseases have no effective treatment. It is usually simpler and safer to destroy a few sick animals than it is to treat them and risk spreading the infection. This is particularly true of animals that have respiratory infections.

PREVENTING SORE DEWLAP

The dewlap, or fold of skin under the rabbit's chin, may become sore during warm weather. The usual cause is frequent drinking from water crocks. The constant wetting causes the fur on the dewlap to become green and foul and the skin on the dewlap and the inside of the front legs to become irritated and rough.

Remove the cause, or, better, prevent this from happening in the first place by placing a board or brick under the water crock. Raise the crock high enough that the dewlap doesn't get wet when the rabbit drinks.

PREVENTING INJURIES

Rabbits are often injured by improper handling. Never lift rabbits by the ears or legs. To carry a small rabbit, grasp the loin region gently and firmly. Put the heel of your hand toward the tail of the animal. To carry a medium or large rabbit, grasp the fold of skin over the shoulder with your right hand and support the rabbit by placing your left hand under the rump. If the rabbit scratches or struggles, hold it snugly under your left arm.

Toenails of rabbits confined in hutches do not wear normally. The long toenails may get caught in wire-mesh flooring and the rabbits then injure themselves trying to get loose. You can prevent this problem (especially in your mature breeding stock) by cutting the toenails periodically. Hold the rabbit's foot in good light and observe the cone in the toenail. Use side-cutting pliers and cut the toenail below the tip of the cone. When done correctly, this operation will not cause hemorrhaging or injury to the sensitive part of the toenail.

To retard growth, cut when both the sun and moon are in barren signs. A decreasing moon and the fourth quarter are preferred. Barren signs are Leo, Aries, Virgo, Aquarius, Gemini, and Sagittarius, which are listed according to the degree of nonproductive qualities.

AT HOME WITH A RABBIT

Though they are not as responsive as dogs, as lively as hamsters, or as quiet as mice, pet rabbits do make very cuddly, comforting animals to have around. If you have enough patience, you can housebreak rabbits to a litter box. Your rabbits can be given the run of the house instead of being confined to a hutch, although it's a good idea to keep the hutch on hand for sleeping quarters or for unsupervised gnawing and burrowing (furniture legs and nice thick rugs can be very appealing). Otherwise, rabbits are no more troublesome than cats.

BIOLOGICAL RHYTHMS

Every plant, animal, fish, and insect responds to external forces — for the most part, forces unseen and unknown.

Llewellyn George believed the influence of the moon was of primary importance: "It is not the light given or reflected by the Moon, its quantity or intensity that we are concerned with, it is the quality of the rays which the Moon transmits which is so significant of its influence according to astrology."

Humans, too, are influenced by many biological cycles. Mood, temperature, blood pressure — all rise and fall at regular intervals as do accident-proneness, intellectual force, emotional attitude, and physical energy. These cycles are variable, starting at birth when the human entity begins to function independently of its parent.

It has now been established that most animals, if not all, are able to "tune in" to Earth's magnetic field.

CHAPTER 6

THE SIGN OF LEO FOR "LITTLE LIONS"

No cat is ever quite like any other cat. No matter how many enter your life, each will have a distinct personality.

Black Stuff cannonballed into our lives one cold November morning just as we were leaving the house to attend church. Running a bit late, we were all in a hurry, and Carl had gone out to start the car. Just as the children and I came out the front door, we looked up, suddenly aware of a black shape streaking toward us down the street. Why that cat was traveling so fast I'll never know; nothing was chasing him, but he sure was in a hurry. He catapulted through the hedge and landed at my feet in nothing flat.

He looked up at me and I looked down at him. It was instant friendship. I had no time to feed the little black devil but I did stop long enough to pet and talk to him for a few minutes, remarking to my daughter as I did so that I hoped he would stay around until we returned.

I need not have worried: We were to be cheered, entertained, and frustrated by his presence for the next 11 years.

When we returned an hour or so later, there he sat on the front porch, waiting. I took him into the house and fed him some warm milk, which he accepted with a lordly air, after which he cleaned up the breakfast scraps. A nap in front of the fireplace (on my best hooked rug, of course) came next. That matter taken care of, he unfolded the rest of his program, most of which seemed to consist of following me about. I have never been honored with so much affection from any animal before or since. I was to find out in the years ahead that he dearly loved being carried and would often suggest this by trying to climb up into my arms.

I simply could not believe that an animal so thoroughly domesticated did not belong to someone. Duly, in the weeks that followed, I made inquiries throughout the neighborhood. I had a dual purpose in asking around. Hoping to keep the cat as a family member, I wanted to take him to our vet for rabies and distemper shots as soon as possible. When it developed that no one claimed him, I did this and breathed easier. This vaccination for dogs and cats is something I feel is never wise to neglect.

At the time that Black Stuff adopted us, both my husband and I were working and had to be away from home during the day. The children were in school. We decided to let him remain an "outdoor" cat and subsequently brought him indoors at night only in the very coldest weather. We arranged a bed for him in a sheltered place by our back door. Carl made a comfortable box out of scrap lumber and I put an old blanket in it, placing it over leaves from our black walnut tree. (These were to drive out fleas, and were replaced from time to time.)

Black Stuff in his box

As Black Stuff matured, we discovered we had a real treasure. In a short time the rats and mice that had plagued our storage feed area for our chickens and stock began to disappear. Some cats are better than others in this respect. Well, this cat was a natural. He even extended his activities to killing snakes. He would bite them neatly in two, right in the middle. Mostly they were of the harmless variety, but there was to come a day when his experience would pay off, his antagonist a deadly copperhead.

We were living on the edge of town then. Meadows and fields surrounded us, and abounded in all sorts of wildlife. A low-lying brushy area a short distance from our property was a known breeding place for copperheads, and they occasionally found their way into our garden. After the day my son came in and asked me to come out and look at the "big worm," I kept a constant watch. I found my best weapon to be quite simple, and usually close at hand: a hoe. When the Lord spoke to the serpent and said, "I shall put enmity between thee and the woman," I'm quite certain he meant me. There is just nothing I like less than a snake, though I have learned, in time, to recognize the harmless and often helpful ones and let them live.

Black Stuff was my constant companion. When I was out of doors working in the garden, he was never far away. It happened that one morning I decided to work on a thickly weeded patch overgrown from neglect. Black Stuff watched as I patiently pulled and dug. Suddenly and without warning he jumped up and spat at me. This was absolutely unprecedented. Taken by surprise, I jumped back, and probably just in time. Black Stuff, who was lightning quick, pounced into the weeds where I had just been about to put my hands. A fight ensued, the likes of which I had never seen.

The copperhead he had by the back of the neck was almost as big as he was. They thrashed about in the high weeds, the snake curling and uncurling about his body. Black Stuff held on and emerged triumphant. And then, as the victor, he brought the spoils of war to me. (This was another strange habit of his and I often suspected Siamese somewhere in his ancestry, for many cats of Siamese breeding are known retrievers.)

BRINGING HOME BABY

I later had a Siamese in which this retrieving trait was pronounced. This delightful little queen (female cat) was the gift of a friend who bred animals for sale and show. Taking my advice, the Queen Mother had been bred so the litter would be brought 63 days later in the sign of Libra. Outdoor cats, particularly American shorthairs, intended to lead useful lives as ratters and mousers, may be bred in Taurus, the sign of strength as well as voice.

My friend was delighted when her queen brought a litter of four beautiful kittens, and so was I when she told me I could have one at weaning age. At 10 weeks of age we brought our treasure home. She was still so tiny that we called her Baby.

Somehow the name stuck, and we never did call her anything else, though on her registry papers she has the high-sounding title of Princess Kaw-lin.

Of course it was unthinkable that Baby be put outdoors, subject to all sorts of danger such as roaming dogs or passing cars. She was kept indoors from the start, and a little box was installed for her convenience. According to your pocketbook and inclination, this can be filled with sand (which we could pick up free from a creek bed), or a commercial material with a built-in deodorizer (which you can buy at most grocery or feed stores). Change the sand daily; the commercial material, under most circumstances, can be used for about a week.

We never used the sand from the cat's litter box either in the garden or on the compost heap. The feces of the cat may contain the tiny crescent-shaped parasite called *Toxoplasma gondii,* which causes a puzzling parasitic disease, toxoplasmosis.

BABY GROWS UP

A female cat matures more rapidly than a tom, usually coming into breeding age at about 7 months, although she may reach breeding age as early as 5 months.

She is capable, during her lifetime, of producing dozens of kittens and may do so well into old age. Unless you wish to breed animals for sale, you should seriously consider having your female cat spayed. In the care of a good veterinarian, this is a relatively simple, safe operation. Neutered cats are clean, quiet, and, if properly fed, will not put on excessive weight. Choose a date when the moon is in Capricorn, Aquarius, or Pisces.

Many veterinarians believe the best age for spaying is 4 to 5 months, before the female comes into her first season. We wanted Baby to have one litter before this was done and consulted our vet, who told us the operation could be safely performed at a later age. He also told us it would be best not to mate her until she was a year old. And we were glad to find out that he had a fine registered tom whose services could be engaged.

Baby came into her first season when she was 5 months old. Well fed and well grown out, she had matured a bit more rapidly than we had expected, and this posed a problem.

Although the female cat goes into heat many times during the year, she usually has only a few peak periods, called estrus, in which pregnancy can occur. Generally there are two, but sometimes three, such periods. They usually occur one in the late fall or early winter and the other as winter turns into spring. Each lasts about 15 to 21 days. The moon and the fifth house rule conception, fertilization, and the period of gestation.

Your female cat's readiness to mate is indicated by the heat period, which usually occurs in three stages. Initially, as her reproductive system prepares for mating and pregnancy, the vulva begins to swell. Her appetite will likely increase and she becomes restless. At this stage, she is attractive to male cats but will most likely refuse their attentions. Don't allow her to escape outside if you want purebred kittens.

It is during the second stage that mating usually occurs. The vulvar swelling continues, now quite apparent. Her attitude becomes more affectionate, her appetite decreases, and she will call persistently for a mate. In the Siamese this cry is quite piteous. For Baby, and for us, it was a bit of an ordeal. Black Stuff, necessarily banished, suffered as well.

The next time a heat period occurred we were more experienced and better able to handle the situation. In time she was duly mated to the Siamese tom, a real beauty. Since Baby was the result of a Libra mating, we chose to have her bring her first litter in a Cancerian sign, hoping for an easy birth and many kittens.

MAMA BABY

Baby, in perfect health, presented no problems at all during her pregnancy and she and Black Stuff resumed their friendship. As with the tom before her, she had been inoculated at the proper age against infectious feline enteritis (distemper) and rabies. This is extremely important, as temporary resistance to the disease is passed on to the kittens. Without it they could die before they obtain protection from their own inoculations. Get your vet's advice on this.

A pregnant cat does not need any special attention. Until the last week she is perfectly able to run about and play as well as ever. During the last week, however, keep her quiet, if you can, and never allow her to become wet or chilled. However, exercise on a slowed-down basis is good for her. Our children understood this and cooperated by playing quietly and handling her gently.

For a maternity box I chose a corrugated carton in which I placed an old bath towel made soft by many washings. I made the side entrance 4 inches above the bottom so that it would be easy for Baby to step over and still keep the kittens in. The box was large enough to let her stretch out full length on her side and have room to spare at both head and tail. My box had no top, so I draped it securely with an old blanket to darken the inside. Conscientiously, I placed the box in a quiet, warm, draft-free site, making sure it was out of the way of the family's usual traffic patterns.

I did everything just right, according to the book, and Baby would have none of it. With complete lack of concern for my convenience, she decided to have her kittens on a bitter cold January night at 11 o'clock. And she determined to have them at the foot of our bed. Why not? She had always slept there, and that was her place.

As I became fully awake and aware that Things Were Starting To Happen, I picked her up gently, placed her in the nest box, and carried her, box and all, into the living room, where the logs in the fireplace still glowed. I put another on the embers, figuring this was going to be a long night.

When I returned to the divan on which I had placed the nest box and sat down beside her, she promptly came out of it and climbed onto my lap. I decided to compromise. Once more I returned her to the box and placed box, cat and all on my lap. This apparently satisfied her, and she then settled down to business.

For a first kittening she had a very easy time, shelling out kittens like peas out of a pod, and she didn't spill a drop of blood. The kittens came seemingly without effort, sliding out wet and perfect. She would cry a little each time and then resume her purring as I petted and talked to her until the next one came. Five times this was repeated, an unusually large first litter. The whole business was accompished in less than two hours. And the kittens were large and beautiful.

When it was all over, Baby was quite willing to stay in the box, so I gently placed it at the end of the divan and covered it with the warm blanket, weighting the overhanging corners so it would not fall on top of her. Baby had done everything according to Hoyle, tearing the sac around the kittens, nipping the umbilical cord with her teeth, and eating the afterbirth as is normal. She had also cleaned the kittens' nostrils of mucus with her tongue and licked each youngster from stem to stern, drying it off. (When a cat does this, don't be surprised if she seems to rough them up a little. She is doing as nature bids her, stimulating the kitten's circulatory and respiratory systems.)

There were moments when I wanted to help, but I resisted the urge. Baby had everything under control. Remember this, and leave your cat alone unless you see a kitten really is in danger. Sometimes a new mom may accidentially lie on a kitten when another is being born; then gently move it.

If you are present during the delivery, make sure that all the afterbirths have been expelled; the retention of one may cause difficulty for the new mother. Afterward, there will usually be a dark red discharge. Bright red may indicate danger of hemorrhage. A greenish discharge can mean that an afterbirth has been retained or the possibility of infection. In either case, consult your vet without delay.

After the birth, you should leave the mother completely alone and let her rest. I knew this but decided Baby was probably thirsty, so I warmed some milk and held it for her to drink, which she did gratefully. This is not necessary and may even be inadvisable with a nervous mother cat, but Baby and I had always understood each other.

For the next few days I fed Baby in the nest box, with her kittens lying serenely beside her. I knew she needed to nurse them frequently. She came out only for brief intervals to use her litter tray.

BABY'S KITTENS

Kittens are born with their eyes closed and remain unable to see for about 10 days. During this time, keep them in a dark room. And you don't have to worry about them learning to eat. They will crawl naturally to the source of dinner, even though both blind and deaf.

On the third day, clean the nest box, and by this time, the mother cat will be glad to have you do this. She is a tidy housekeeper and keeps the nursery in good order by washing the kittens and disposing of their waste.

Baby's kittens were awaited eagerly by others besides our family, for we had promised two to friends; we had decided to keep the three little toms.

I didn't let Black Stuff into the house until a day or two after the kittens were born, but when a snowstorm developed I couldn't bear to keep him away from the warm fire any longer. Delighted to have his privileges again, he raced into the living room. Knowing this would again be a critical time, I stayed near. Tom cats sometimes eat kittens, and I wasn't sure what to expect.

What happened was ludicrous. As Black Stuff passed by the nest box, one of the tiny kittens mewed. In amazement, our big brave cat made for the nearest chair, slunk under it, and stared out in terror. Black Stuff, who fought everything with gusto, had been scared out of his wits by a tiny kitten.

By now Baby was alerted to his presence and jumped down to greet him, just as she always did. Reassured, he came out but remained wary. This soon wore off, and he lay down in front of the fireplace and went to snoozing.

Figuring all was well, I went back to my household tasks. It wasn't long after this that I heard Baby scratching in her litter box. When I returned to the living room to investigate, Black Stuff was lying in the nest box serenely permitting the kittens to tumble around him. I was sure then that he had Siamese in his ancestry; this attitude is typical of the highly intelligent tom of this breed.

During the time that Baby nursed the kittens, she was permitted to have all the food she wanted. This is important: The mother cat must supply energy for her own body to return to normal and at the same time give the kittens the milk they need during their early growing period.

Don't rely on table scraps to feed your cat. They are a good supplement to the cat's diet, and cats enjoy the variety, but for the sake of cleanliness and economy, we have always preferred to feed our cats a dry cat food. There is little or no waste and no danger of spoilage. However, when Baby was nursing, she was allowed free choice of the dry food or the semimoist canned product.

One morning I went into the living room as usual and found the nest box empty. True to her instincts, Baby had hidden her kittens. I found them in the back of a dark closet and let them be. Still Baby wasn't quite satisfied. She moved them again several times. One night she brought them in, one by one, and placed them all at the foot of the bed, where she was accustomed to sleeping herself. I moved them but when she returned them the second time I gave up. The next morning she moved them again.

She finally ended up by putting them in my office under the bookcase and seemed content with this location. I removed the throw rugs and put down an old plastic tablecloth. By now it was necessary, since kittens are not born housebroken.

Baby moving her kittens

BABY TRAINS HER KITTENS

I brought in Baby's litter box, filled with fresh sand, and she set about teaching her kittens to use it. She picked them up from time to time, held them a few inches above the sand, and casually dumped them in. Instinctively the little guys would dig and deposit. In just a few days, all five kittens were using the box, neat as a pin.

I'm not exactly sure when they began eating dry cat food. I would simply fill Baby's dish any time I saw it empty, but after a while I began to notice that it was getting empty more often than usual. Kittens are great imitators, and I'm sure that when they observed their mother eating, they became curious and decided to try the new food too. Kittens have well-developed teeth and jaws and are quite able to eat dry food.

TIME TO SAY GOOD-BYE

When Baby's kittens were 10 weeks old, the two little females went to new homes. Leo, Theo, and Nero, the toms, remained with us. Baby was a smart cat but fortunately she couldn't count. After a brief period of searching, she settled down to comfort herself with her sons and didn't seem overly upset. The kittens were well past weaning age, of course, but we felt the parting would be easier if they were removed in the sign of Capricorn, and this proved to be the case. Sagittarius, Aquarius, and Pisces are also good signs.

CATS AND HERBS

Basically cats are carnivorous (meat eating) and need a high-protein, high-fat diet. Meat, fish, fowl, and certain vegetables such as soybean provide the nutrients they need. In the wild state they kill their prey and usually devour the entire carcass, including whatever greens may have been eaten. By doing this they achieve nutritional balance. Domesticated cats, especially those kept in the house most of the time, do not have this opportunity.

I always grow catnip for my cats. It's well known as the "cat herb" and is tonic for them. I also dry catnip for winter use.

Another herb attractive to cats is **valerian.** The root is the part they like.

If you have both "indoor" and "outdoor" cats, fleas may be a problem. The leaves of **black walnut** are effective against fleas (which the outdoor cat usually brings in). Place the leaves under furniture or rugs. Fresh or dried sprigs of *pennyroyal* have long been used as a flea repellent. The oil is far stronger and more lasting, but you can put the dried herb in pet pillows. If fleas get into your carpets, dust them with rotenone powder.

Chamomile flowers and dried *winter savory* also possess remarkable properties for repelling fleas. (I don't use flea sprays on cats because they lick their fur.)

Cats intensely dislike *rue,* and you can use this knowledge to your advantage. Cats don't scratch at your furniture because of a mean desire to tear it up. They do it because they're annoyed by the ragged edges of their claws. Outdoor cats remove these as nature intended, by scratching at trees or posts. Sometimes they come off in a fight. Indoor cats don't have this opportunity, so they claw anything handy, usually your best chair. Cats don't understand punishment, so slapping at them will only make them afraid of you. Instead, be wily. Always provide your indoor cats with a good, sturdy, scratching post. This can be as simple as a log of wood laid on the floor. We have an upstanding post, covered with carpet, nailed to a firm base. The kittens were introduced to this and urged to use it. They seldom fall from grace, but when they do I rub the furniture with rue and my chairs are safe again.

I try to let the Siamese outdoors, in the exercise yard as often as possible, and one reason is to give them the opportunity to eat grass. Cats are almost constantly licking themselves and each other and in consequence swallow a great deal of fur, most of which simply passes through their system. However, as with dogs, who eat couch or twitch grass to promote cleansing vomiting, cats also like to partake of a blade or two from time to time. They seem to prefer the slender, rather tough stems of Bermuda. A short time afterward, they upchuck, relieving their systems of any undigestibles that are bothering them.

TOYS

Most of us like to have cats around because of the keen enjoyment we derive from their presence. Provide your cat with some toys or he may find some for himself.

Cats will play intermittently for hours with a spool attached to a strong string and hung from a doorknob. We would take a sturdy box and cut out a door in one end and a window in the other. When one cat crawled inside, the others stayed out. They'd slap at each other playfully through the door and window. Cats also love to play with a paper bag or a shoe box, a rattling ball, a rubber mouse that squeaks, a catnip toy, or pipe cleaners curled into a spring.

Outdoor cats often frolic with leaves that are blowing in the wind; indoor cats get much the same pleasure from a piece of crumpled cellophane, enjoying the noise it makes. Our cats also love to crawl under newspapers, pretending to hide while the others jump on the moving paper.

Caution: Don't let your cat have furry, rubber, or woolen toys that he might tear apart and swallow. Toys with small or sharp parts such as buttons, beads, and rubber bands may also be troublesome. And be careful of plastic bags such as those that come from the cleaners. Don't leave one about that your cat can crawl in.

We sometimes give our cats a bone with rounded edges to play with or to chew on, but *never* chicken bones (which splinter) or one with a sharp edge.

Don't give your cat too many toys at a time. Change them occasionally and put the "old" ones away. Then, when you bring them out in a few weeks, he will be intrigued all over again.

CAT PROBLEMS

Have your cats inoculated for rabies and feline enteritis (distemper). Consult your veterinarian about any other vaccinations that are required or recommended.

If you suspect that your cat has **worms,** indicated by a dull coat, inflamed eyes, coughing, and vomiting, you may find that garlic tablets (also good for dogs), are an effective remedy.

Cats that hunt for and kill mice sometimes get **tapeworms,** since they are carried by rodents (as well as birds, bats, fresh-water fish, and amphibians). Keeping your cat's quarters clean and free of fleas is important. If your cat eats fleas, the tapeworm eggs can hatch inside. The cat may give evidence of tapeworm infestation by dragging his rear end over the floor or ground. There is also general loss of good body condition.

Ticks are eight-legged, hard-shelled arachnids, similar in appearance to spiders. They burrow into the skin of cats that are allowed outdoors. They feed on their blood, causing animals to become anemic. Loosen ticks from your cat's skin by soaking them in vinegar, or cover ticks with an oily or greasy substance, which stops the tick from breathing. Use vaseline, nail polish, or alcohol.

Ear mites may be the culprit if your cat starts shaking his head, pawing at his ears, or rubbing his head against furniture. The ears of a cat are quite fragile and the condition, if untreated, may result in ear infection or even deafness. Consult your veterinarian if you suspect their presence.

Constipation can be caused by **hairballs.** Eating grass helps cats to dislodge these accumulations. Proper grooming is also a preventative. Our cats are brushed daily. They love this ritual and would even line up for it every morning, climbing all over Carl, the brusher.

Older cats are more apt to be constipated than are younger ones. Try an occasional teaspoon of salad oil or butter or a meal of raw liver. You can also use mineral oil, but don't give more than an eyedropperful twice a week.

When disease in cats is suspected, perhaps the most important thing is to be able to recognize the condition, or at least the fact that the cat is ill, and promptly get help. Then follow your veterinarian's advice. Feed your cat only the prescribed medication. Never give your cat aspirin except on your vet's orders; it is very harmful to a cat's stomach.

> **BUYING A CAT?**
> Buy cats or kittens when the moon is not aspected with Mercury.

If your cat is sick or injured and cannot wash himself, help him to keep clean with a damp washcloth and a small amount of mild soap. Do not wet the cat, but just sponge his fur with the moistened cloth. If the animal is injured and unable to move, turn him over several times a day to prevent sores.

BEWARE THE PLANTS

A problem new cat owners may not be prepared for is the animal's persistent but potentially dangerous habit of chewing on plants. Some cats are only occasional chewers when they are bored or frustrated, but others are obsessed with the idea of destroying every bit of greenery in the house.

Some common houseplants, including the wandering Jew and the spider plant, are not harmful, but many are: Jerusalem cherry, poinsettia, philodendron, elephant ear, mistletoe, oleander, azalea, and English ivy, for example, as well as a number of garden plants, such as lily of the valley, foxglove, delphinium, privet, boxwood, yew, and larkspur.

Be sure to keep all potentially harmful plants out of reach; check with your local arboretum, the Department of Agriculture, or reference books if you are not sure about the toxic nature of any plants you have.

> **A DANGEROUS MYTH**
> It is simply not true that "a cat always lands on his feet." A cat, dropped or falling from a height, can be severely injured. Don't experiment!

CHAPTER 7

SIRIUS, THE DOG STAR

The taming of the dog marked one of the great events
in the progress of civilization.

Many of the constellations bear little or no resemblance to the names that have been given them, but Sirius — or Sothis — whose heliacal ("relating to the sun") rising heralded the first day of the Egyptian new year, was so named simply because this particular group of stars looks like a dog.

Sirius, or the Dog Star, is the brightest star in the heavens. It can be seen by looking along an imaginary line pointing southward through Orion's belt. Sirius is the head of the constellation Canis Major, the Great Dog.

The ancient Greeks originated the phrase "dog days" for the hot, sticky days of summer, applying it to the period when the Dog Star, Sirius, rose with the sun. This is a period of about 40 days beginning in early July and ending near mid-August.

The rising of Sirius does not actually affect the weather. However, the middle part of the 40-day period happens to occur at the same time as most of the uncomfortably hot days. Not so long ago some people thought that "dog days" were so named because dogs were most likely to contract rabies during hot weather. We now know that this is not true, and the superstition has largely disappeared.

* 100 *

MAN'S "BEST FRIEND" IS ANCIENT

Dogs are descended from an animal called Tomarctus, which lived about 15 million years ago. Its descendants developed into wolves, jackals, coyotes, foxes, and other wild dogs that spread throughout the world.

After humans tamed dogs, various breeds were developed for specific purposes and gradually became fixed in size, color, and the ability to perform certain tasks such as hunting, herding, guarding, or serving as companions. Breeds were named for the game they hunted, the work they did, or the places where they were developed.

Dogs are ruled by Mercury. If you contemplate buying a dog, do so when the moon is not aspected with the sun.

ACQUIRING YOUR DOG

Before you get a dog, think the matter through. Decide whether he will be kept in the house or the backyard, or given free range.

Find out what the laws are regarding dogs in your area. Most cities and states have statutes that require licensing. Many communities require dogs to be leashed or penned when outdoors, and some require that penned dogs be kept quiet at night to prevent disturbing neighbors.

We kept our dachshunds quiet by penning them in a large, comfortable doghouse. If they barked, they were slapped a few times with a folded newspaper, which does not hurt but makes a lot of noise. However, farmers or ranchers living in isolated areas where dogs are allowed to run free often find their barking beneficial as a warning of intruders and may even train them to bark loudly when strangers approach.

With these things in mind, decide first of all whether you want a small dog or a large one, a hunting dog, a shepherd dog, or perhaps just a dog as a pet for a child or for you. Do you want a purebred or will you be just as happy with a mixed breed?

HOUSING YOUR DOG

All dogs do better when they are allowed to live and grow up in the house. They are happier, have better manners, and learn faster than dogs that are kept tied in the yard. For many reasons, though, this is not always possible.

If you must keep your dog outside, arrange for a dog pen and doghouse before getting an animal. A doghouse need not be elaborate and you can build one yourself out of scrap lumber. It should be weatherproof, however, keeping out rain and wind. An old blanket gives added warmth during the winter months.

Doghouse-building plans are easily obtainable, but the main points are these:

- ✿ Place the kennel off the ground sufficiently to prevent water from seeping in.
- ✿ Do not make it too large; otherwise, the dog will be unable to heat it in the winter.
- ✿ There should be an overhang at the door to keep rain from coming in. The roof should be sloping.

THE KENNEL RUN

A dog kept outside needs a place where he can exercise. A run 10 feet wide by 20 feet long is adequate for most dogs. Galvanized steel woven into what is commonly called storm fencing is best. Six feet is an average height sufficient to prevent the dog from climbing over. If this is too expensive, use a heavy grade of chicken wire (1-inch mesh or less), provided it is put up securely.

When building your run, be sure to set the posts when the moon is in the third or fourth quarter and in one of the fixed signs, Taurus, Leo, or Aquarius.

Paved runs are expensive and also hard on a dog's feet. Besides, I feel that dogs, like other animals, benefit from contact with the vital radiations of soil. For the sake of cleanliness, a compromise is to use medium-size gravel. If the run can be located on a slight slope, which would permit good drainage, by all means place it there.

Shoo, Flies

If flies in the run are a problem, spray or sprinkle with the following mixture: 3 parts oil of cloves, 5 parts oil of bay, 5 parts oil of eucalyptus, 150 parts alcohol, 200 parts water. Diatomaceous earth sprinkled in the run will also help keep down both flies and odors.

IF YOU CAN'T HAVE A PEN

For various reasons, including expense, it may not be possible to have a fenced pen, or perhaps you need some other means of exercising your dog until the pen can be built. In such a situation, consider a wire run. String wire from one well-set post to another or from the back porch to a tree. Put a ring on the wire and fasten a leash to this. Fasten the other end of the leash to the dog's collar. This arrangement permits the dog to run from one end of the wire to the other. He may even be taught to relieve himself at one end of the run.

Dogs kept in the house need exercise as well. Leash the dog and take it for an escorted walk. Plan for a regular schedule of walks and playtime.

Exercise requirements will vary, of course, as will those for being taken outdoors for relief. The latter depends on how many times a day you feed, how much, and how much water the dog drinks.

CHECK OUT YOUR DOG

Before accepting a gift dog, or buying one, or getting one from the pound, there are some important matters you should check on. Find out whether the dog has been wormed and how long ago. Were worms found? If so, what kind?

It is also necessary to know whether the dog has had protection from rabies, distemper, and hepatitis. The age of the dog must be taken into consideration here; some breeders feel that puppies enjoy immunity as long as they are receiving their mother's milk. If the pups have just been weaned, they may not have had their shots. Even if the pup has had serum protection, you should speak to your veterinarian about active protection.

BUYING A PUREBRED?

Get your registration papers at the time of purchase. Your dealer should be able to give you a three-generation copy of the puppy's pedigree and an application form to register a puppy from a registered litter. The dog's registration will then be signed over to you. Have the papers transferred so ownership is in your name. I repeat, *get the papers at the time you get your puppy.* If they are not available, have the dealer (or breeder) give a you a written statement that you will receive the papers within a definite time or your money will be refunded.

At the time, this may not seem important, but later you may want to show your dog or compete in obedience trials. If you purchase a female, you may want to breed her when she is mature (and possibly sell the pups). If your dog is a male, you may have requests from owners of females to breed to him (and you can also charge a fee for this).

Once you have received your registration papers, send them away at once to the registering body so they won't get lost and you won't forget.

A SOOTHING CLOCK OR FROCK

A new puppy may be troublesome at first, howling and whining at night because it is lonely and in new and strange surroundings. Place a ticking clock next to his box to keep him quiet. Lacking this, give him an old garment of yours that has not been washed.

FEEDING YOUR DOG

Dogs are carnivorous by nature; that is, the natural canine diet is raw meat. Their internal organs are equipped with very strong digestive juices, and they have strong, muscular stomachs and intestines. These muscles must be exercised, and for this reason raw meat is better than cooked meat, which is already semidigested.

Dogs in the wild are scavengers. They prey on sick and old animals and will eat decomposing carcasses. (The disease bacteria are destroyed during the dog's normal process of digestion.) It is a dog's normal instinct to bury a bone, digging it up again when the process of decomposition has begun. You can feed your dog cereals, but he will have greater difficulty in digesting these rather than flesh foods and will need them in greater quantity.

All of our dogs have always loved fruit, avidly eating apples and pears from our small orchard and enjoying many discarded vegetables.

With today's high prices of meat, table scraps are apt to be scanty. Even organ meats are expensive, and some shops now charge for bones. We must always adjust to the framework of the possible. And the possible, in this case a commercially prepared food, is not at all bad. The choices are numerous, and most of the leading brands are quite good. Many authorities don't think you should feed table scraps at all, believing instead that the best commercial foods, dry, moist, or canned, contain more than adequate amounts of vitamins and minerals for normal growth and reproduction.

RULES FOR FEEDING

The quantity fed depends on both you and your dog. Each dog is an individual; even two littermates may differ in their nutritional requirements. For feeding mature dogs, here is a good rule to follow: ⅓ to ½ ounce of dry food per pound of dog per day. Thus, a dog weighing 32 pounds would receive 12 to 16 ounces.

Remember, if you are feeding a canned food, that it contains about 78 percent water. With this in mind, feed 1 to 1½ ounces of canned food per pound of dog per day.

If you are feeding a puppy on dry dog food that has been moistened, keep some before him at all times. An adult dog, on the other hand, should have only the amount he can clean up in 30 minutes.

TIME TO FEED

A mature dog does well on just one meal a day unless he is working hard, or unless you are feeding a female nursing young. Too much food will just make him fat or go to waste.

The general rule is to feed dogs in the evening before your own meal. This will keep them from begging. However, you may feed dogs in the morning instead, particularly if you have housebreaking troubles. If noon is your most convenient time, by all means feed then. It is largely a matter of your own choice but once you pick a particular time, stick to it.

Sometimes it's wise to rest your dog's digestive system through fasting. Enforced fasts are natural to dogs in the wild state, and it is at this time that the accumulated toxins of a meat diet are expelled from the system. Wild dogs cannot always kill with regularity when they are hungry, and a wise Nature has decreed that this time of fasting should be put to good use.

Fast a puppy once each week for half a day. After 4 months of age, fasting should be more frequent. Fast an adult dog regularly one day each week. While he is fasting, give your dog a bit of watered milk with honey.

Your dog should always have fresh water readily available.

THE PREGNANT DOG

Only a small increase in food consumption is needed for a female carrying puppies, but the nutrient quality of the ration is important at this time. Nutrients are being stored in the developing pups, but the female, now less active, is also using her food more efficiently.

The female dog should be bred so that she will produce her pups in one of the fruitful signs, Cancer, Scorpio, or Pisces, in the moon's first quarter. These are the water signs, productive of best growth in all things, whether plant or animal. The second best choice is Taurus or Capricorn, the earth signs. As with other animals, pups born at this time make the best stock for future breeding. The average gestation period is 63 days.

After her pups are born, the mother needs additional feed to enable her to produce the quantity of milk her litter requires. By about the fourth week (with a litter of normal size), she will be consuming about twice her usual amount of food. Normal litters are six to eight pups. With a larger litter, she may need a food supplement such as horse meat or hamburger in addition to her dry rations to enable her to produce sufficient milk. The proportion should be 1 part meat to 3 parts of the dry ration.

WEANING

At about 3 weeks of age, most pups begin eating the same solid food as their mother. By 5 to 6 weeks, the female has usually started to wean the pups. By 6 weeks, most pups can be completely weaned and on solid food.

As with other animals, choose a day when the moon is in Sagittarius, Capricorn, Aquarius, or Pisces. Let the pups nurse for the last time when Venus is well marked.

On the day when weaning takes place, do not feed the mother at all. On the second day, feed one-fifth the normal ration, and on the third day two-fifths. Gradually add some each day until she is back to her normal ration. This helps to restore her milk-producing organs to normal by lessening the quantity of milk she produces.

FEEDING THE PUPS

Keep an eye on things and make sure that all the pups get a chance to nurse. Often, the stronger ones push the weaker pups aside. If you make the mother's ration available to the pups, they will probably be eating enough by 6 weeks of age so that no setback occurs at weaning time.

Keep a good puppy chow available at all times until the pups are 16 to 20 weeks. After that, one feeding a day should suffice. However, if the pups do not overeat, self-feeding can be continued. Just take care that a pup does not become overweight, which is detrimental to general health and bone development.

A pup's energy needs are roughly twice that of an adult dog. The ration should be high in protein and fat and contain all essential vitamins and minerals.

Where you will keep your pup also makes a difference: Those living in unheated kennels during winter weather need more food than those in warm kennels. Puppies living in a house (or those difficult to housebreak) can be fed once a day.

LEARNING TO OBEY

The ability of dogs to learn to obey commands is a mark of their intelligence. A puppy, for the first 3 weeks of life, knows nothing and needs only warmth and food. In the fourth week it becomes more alert to its surroundings. By now it can see, hear, and smell, and it begins to learn. From the 4th to the 7th week it starts to play and respond to humans. From the 7th to the 12th week, it can learn simple commands such as "Let's take a walk." Repeated several times, such orders will come to mean to the dog exactly what they mean to a person.

Make full use of your dog's intelligence, but don't overestimate it. Many times dogs are punished for disobedience by owners who have failed to make them understand just what they should and should not do.

Never shout at your dog; teach him to obey commands in a quiet voice. Never try to make him do something when you know perfectly well he won't. An order that is obeyed is a step forward; one that is not carried out is a step back. One of the first essentials in training is to avoid giving the dog the chance to disobey.

There is no need to be stern, but you must be firm. In fact, if you can turn your dog's training into one big game, so much the better. Just bear in mind that it is you and not the dog who must control the game, and it is yours to say when the game is over.

HOUSEBREAKING

This should start as soon as you bring your puppy home. Housebreaking depends on the millions-of-years-old desire of den dwellers to keep their beds clean. You can make use of this ancient instinct.

Paper training a puppy

If the puppy is to stay indoors, place newspapers on the floor of the kitchen or another room that will not be easily damaged. When the puppy wets, rush it to the papers, which should always be placed in the same spot.

Be sure to get the puppy to its papers immediately after it eats, plays, or wakes up. Always praise the pup when it wets where you want it to. Scold it when it makes a mistake, but do not rub its nose in the droppings.

Use this same method when you want your dog to "go" outdoors. Watch where the puppy wets and then always lead him to these places when you take him outside. Or select a spot yourself by mopping up his first mistake and anchoring the rag to that spot. The puppy smells where he has been before, or thinks he has, and will seek out the same spot again.

Take your puppy to this "spot" the first thing in the morning and the last thing at night. Teaching your dog to go outside to relieve himself is preferable to teaching him to go on paper, but whichever suits your lifestyle, start him out that way from the very beginning and don't change it. It is very difficult to retrain a dog to relieve himself outside if he has been started on papers indoors.

TRAINING YOUR DOG

If you establish rudimentary obedience training when your dog is a puppy, training the older dog will be easy. It begins with teaching your dog to walk on a leash, to come when called, to sit, to lie down, and to come to heel. Many good books on dog training are available, or you may want to join a local training class.

You can start this training as early as 8 weeks. Let the sessions be short, not more than 15 minutes at a time, but several times a day. Speak in a quiet voice, and pat and reward when the animal does something right. Make sure the first lesson is well learned before you begin the next one. Always be patient but firm. A dog should understand that a command means instant obedience.

FOR A HEALTHY, HAPPY DOG

Maintain your pet's health with proper food and exercise, but if you feel he is truly sick, take him to a veterinarian. Don't try to dose him yourself if you suspect a serious illness.

The nursing puppy usually receives protective antibodies from its mother's milk, but this immunity begins to disappear when the puppy is weaned and will generally be gone in about two weeks. Consult your veterinarian as to the age when a vaccination program should be started.

Give the in-whelp bitch garlic tablets throughout pregnancy to help keep her worm-free. You can continue this treatment through the nursing period. With the addition of other fresh, finely cut herbs to her diet, the pups will also be free of worms.

If your dog has lice, fleas, or other vermin, bathe him, dry him thoroughly, and rub his body with eucalyptus oil and ammonia. (The proportion 1 teaspoon of the oil to 2 of ammonia, mixed in ½ pint of warm water.) Pay special attention to the neck area, earflaps, and the base of the ear, as well as the brisket, back, and base of the tail.

Place walnut leaves under the dog's blanket and change them occasionally to keep down fleas. Sprinkle a pad or pillow for the dog's bed with chamomile flowers or rue to expel fleas. Fresh or dried sprigs of pennyroyal have long been used as a flea repellent. Use herbs, too, to stuff the pillows, adding more occasionally to freshen. Rub oil of pennyroyal on the dog's fur, or use tansy in late summer.

Don't be alarmed if your dog vomits occasionally. As with cats, this is nature's way to effect rapid internal cleansing. Dogs, and puppies as well, seek out and eat couch grass to induce cleansing vomiting. If your dog vomits solid food, fast him for a day to give his digestive organs a chance to rest.

Puppies frequently have roundworms or hookworms; an older dog may have tapeworms. If you suspect their presence, take a stool sample to the veterinarian, who will prepare slides and check them under a microscope. He or she knows the right medicine to prescribe. Worms are most active at the time of the full moon when they are stirring and breeding, and thus easier to dislodge. All worm treatments are best given when the moon is waxing and near the full. They may also be given (though the time is not quite as good) when the moon is just on the wane.

Thymol, an extract from the thyme plant, is helpful for dogs with hookworm. Give 4 to 6 drops daily on sugar lumps. Better results will be achieved if a laxative diet composed of bran, grated raw carrot, and some hamburger is maintained at the same time. Worm your dog at the time of the full moon.

Worms in older dogs, especially farm dogs, can be quite serious. Tapeworm is also a danger to sheep (and it is usually taken up when they are grazing grass where dogs have been). This worm can encyst in the brain of sheep and cause the often fatal ailment gid. Worms are difficult to prevent and get rid of — even a creature as tiny as a flea can serve as the host for tapeworm, and almost all animals have fleas at some time or another.

Dogs like the flavor of garlic, and we have found that mixing ½ to 1 whole garlic tablet with our dogs' food is helpful in eradicating worms. Our dogs will eat garlic cloves mixed in with their food and even seem to relish them.

HUNTING WITH YOUR DOG

It is best to begin the training of hunting dogs when the moon is increasing and in good aspect to Jupiter or Venus. Start actually teaching them to hunt when the moon is in Aquarius and in good aspect to Mars and Jupiter. Hunting with dogs will be most successful when the moon is in Gemini, Libra, or Aquarius, in good aspect to Mercury and Venus, and not adverse to Saturn or Mars. Favorable and unfavorable days change each year, and you will find them listed annually in *Moon Sign Book and Gardening Almanac* put out by Llewellyn Publications.

Hunting birds by gun or by hawk is considered best when the moon is in Aries and not adverse to Mars. Hunt birds by trap or with dogs when the moon is in Gemini, Libra, or Aquarius and in good aspect to Mercury and Venus.

CHAPTER 8

THE SQUARE
OF PEGASUS

Wean a foal only when the moon is in Sagittarius, Capricorn, Aquarius, or Pisces. Let it nurse last in a fruitful period.

Early humans did not ride horses; they hunted them and ate their meat. Early horses were, in the beginning, rather small. It is not known who first tamed horses and trained them to be ridden, but excavations at the ancient city of Susa in southwestern Asia indicate that horses were ridden more than 5,000 years ago.

Arabs were the first people to use selective breeding to produce better horses. These horses were highly thought of and carefully guarded, the best of them usually being owned by the higher class of chiefs.

The Arabian has been carefully bred toward a definite type for a longer period than any other breed of livestock. All breeds of light horses and most of the heavy breeds carry to some degree the blood of the Arabs. Therefore, it would seem to make sense to look to the Arab nations for answers to questions of breeding, training, and care.

STABLING AND EQUIPMENT

If you have no stable, check with your county agent for requirements and building plans (as well as information on other equipment needed to keep a horse).

FEEDING HORSES

Horses have smaller digestive tracts than cattle and cannot eat as much roughage. Feeds should not produce surplus body weight or large, paunchy stomachs. If you use small quantities or have little storage room, you may find it more satisfactory to buy ready-mixed feeds.

Good-quality oats and timothy hay are standard feeds for light horses. Hay of more than one kind makes for variety and appetite appeal. In season, good pasture can replace part or all of the hay unless work or training conditions make substitution impractical.

For a balanced ration at lowest cost, interchange feeds of similar nutritive properties as prices go up and down. Some of these feeds are grains (oats, corn, barley, wheat, and sorghum), protein supplements (linseed meal, soybean meal, and cottonseed meal), and hays of many varieties. During the winter months, add a few sliced carrots to the ration, an occasional bran mash, or a small amount of linseed meal.

PASTURE MANAGEMENT

Good pasturage is the cornerstone of successful horse production, and in season there is no finer forage for these animals. A temporary pasture grown in a regular crop rotation is preferable to a permanent pasture that may be infested with parasites.

Since horses are less likely to bloat than are cattle or sheep, legume pastures are excellent for them. Specific grass or grass-legume mixtures vary widely from one area to another according to differences in soil, temperature, and rainfall. Consult your county agent about pasture grasses suitable to your locality.

Horse pastures should be well drained. Make sure that shade, water, and minerals are always available. Pits, stumps, poles, tanks, and places dangerous to horses should be guarded.

Black mustard, a plant of pastures, is cleansing to both animals and the soil itself. It is important as a tonic and as a disinfectant to the earth. The cresslike leaves and seeds, as well as its bright yellow flowers, are bitter tasting, yet animals like to eat it. Mustard promotes appetite, causes salivation, and stimulates the excretion of digestive juices. The seeds are useful as a worm expellent. Use white mustard for the same purpose. The highly antiseptic garlic is also an excellent pasture herb, both for eating and for cleansing the earth. It is particularly good for horses that are infested with worms.

Couch grass, brambles, wild turnip, green fern and broom tops, ash twigs, and elder are all worm expellents. Early-spring grass is one of the best vermifuges for horses. Mares that feed on spring grasses and the shooting leaves of hedges and trees (such as mulberry) provide milk with a vermifugal effect on their suckling foals.

MINERALS

The classical horse ration of grass, grass hay, and farm grains is often deficient in calcium but adequate in phosphorus. Salt, too, is likely to be in short supply, and many horse rations don't contain sufficient iodine. Thus, horses usually need mineral supplements, but don't feed either more or fewer minerals than necessary.

On the average, a horse consumes about 3 ounces of salt daily, or 1⅓ pounds a week, though salt requirements will vary with work and temperature.

The salt requirements, and any calcium or phosphorus needs not met by feeds, can best be supplied by allowing free access to a two-compartment box containing minerals. One compartment should have salt (iodized in iodine-deficient areas); the other should contain a mixture of 2 parts steamed bonemeal and 1 part salt (the salt is added for palatability). It is important to maintain the calcium/phosphorus ratio at not less than 1:1. Where it is obtainable, seaweed is given to horses because of the high iodine content. A good commercial mineral mixture may be used, too.

WATER

Horses need ample quantities of clean, fresh, cool water. They will drink from 10 to 12 gallons daily, the amount depending on the weather, work done, and rations fed.

Free access to water is always desirable; if this is not possible, water your horses at about the same times daily. Most horsemen agree that water may be given before, during, or after feeding. During warm weather, frequent small waterings between feedings are in order; they're also a good idea when the animal is being put to hard use.

Caution: Never allow a horse to drink heavily when he is hot; this may cause him to founder. And do not allow a horse to drink quantities of water just before being put to work.

BREEDING

Horse owners intending to have mares bred and to raise foals should know at least the basic elements of breeding.

A stallion standing for outside service may become the sire of a great number of progeny. Select carefully. He should be a superior individual in type and soundness. He should also be a purebred and a good representative of the breed selected. If it is an older horse that has been used in service, he should have proved to be a prepotent sire, with a great many uniformly high-quality offspring.

The brood mare, in addition to being sound and of good type, should possess an abundance of femininity. And whether purebred or grade, she should be of good ancestry. For buying, choose the new moon or the first quarter of the moon.

Mares generally begin to come in heat when they are 12 to 15 months old. Signs of this are relaxation of external genitals, more frequent urination, teasing of other mares, apparent desire for company, and a slight mucous discharge from the vagina.

Heat periods usually recur at about 21-day intervals. This interval may be as short as 10 days or as long as 37. The duration of the heat period averages 4 to 6 days, but may range from 1 to 37 days. In spring, some mares remain in heat as long as 50 or 60 days.

The best time to breed the mare is in spring, when pastures are green and succulent. During this season mares are generally putting on weight, their heat periods become more evident, and they are more likely to conceive.

There is much to be gained from this plan. A foal conceived in spring may be dropped on pasture the following spring. At this time there will be a minimum danger of becoming infected with parasites or diseases. Abundant fresh air, sunshine, and exercise all contribute to the good development of the youngster.

The average gestation period of mares is 336 days, a little over 11 months. This period may vary as much as 20 to 30 days either way, depending on individual mares.

A handy method to figure the approximate date of foaling is to subtract one month and add two days to the date the mare was bred. Thus, if a mare was bred May 20, she should foal April 22. Use this in figuring ahead to the approximate birth date if you wish the foal to arrive in a particular sign.

AGE TO BREED MARES

Breed mares when they are 3 years old so they will foal when they are 4. Only an exceptionally well-developed filly should be bred as a late 2-year-old. A filly bred as a late 2-year-old should be well fed so that her own immature body, as well as the developing fetus, will grow properly. Give her a rest and don't breed her again the following year.

Brood mares, properly cared for, may produce regularly until they are 14 to 16 years old. In some cases, mares produce until they are 35 or so.

In selecting a brood mare, either obtain one 3 or 4 years old or make certain that the breeding habits of an older mare are regular.

SIGNS FOR BREEDING

Horses have strength, speed, and endurance, but not always in equal amounts. With careful breeding, for instance, a fast horse and a strong horse may produce an animal that is both speedy and powerful.

Strength is emphasized by breeding in Taurus. The fire signs, Aries, Leo, and Sagittarius on the ascendant at birth give the best stamina, and next to these the air signs of Gemini, Libra, and Aquarius. The least stamina is bestowed by the water signs, Cancer and Pisces. Of the earth and water signs, Taurus, Virgo, and Scorpio give good stamina when on the ascendant at birth. The moon has much to do with the constitution at all times. When well aspected at birth by the moon, the sign of Libra influences grace, beauty, and gentleness of disposition.

Mars, rising in Aries or Scorpio and well aspected by the sun, imparts a lively spirit as well as vigor. Mercury, well aspected by the moon at birth and in one of his own signs (as in Gemini or Virgo), gives intelligence.

CONDITIONING THE MARE

Conditioning before breeding is as important for the mare as it is for the stallion, and proper feed and exercise are the keys to correct conditioning.

To achieve high conception rates, mares should not be too thin or too fat. Try to check the natural tendency of barren or maiden mares to become overweight. Overweight animals usually will not breed well.

Condition mares by riding them under saddle or driving them in harness. When this is impractical, they usually get enough exercise if allowed to run in a large pasture.

CARE OF PREGNANT MARES

The pregnant mare is inclined to inactivity. Separate her from barren mares, which are likely to be more frisky and playful than she. The best place to keep her is in a pasture in which she has shade and water. She doesn't need more than a simple shelter. In temperate climates, an open shed is satisfactory even in winter.

The best exercise for a pregnant mare is to be allowed to roam a large pasture. If this is not possible, exercise her for an hour daily under saddle or in harness, before she foals. Do not confine her to a stable or a small, dry lot on idle days.

CARE OF THE MARE AT FOALING TIME

By all means, keep a written record of the date the animal is due to foal, remembering the gestation period can range from 310 to about 370 days (the average is 336). This record will enable you to make plans for the foaling.

Giving birth is a critical time in the life of the mare. The first sign of approaching parturition may be a distended udder two to six weeks before foaling time. Seven to 10 days before foaling there generally is marked shrinking or falling away of the muscular parts of the top of the buttocks near the tail end, and a falling of the abdomen.

Although the udder may have filled out previously, the teats seldom fill out to the ends more than four to six days before foaling; generally there is no wax on the ends of the nipples until two to four days before parturition. The vulva will become full and loose at this time.

As foaling time draws nearer, milk drops from the teats. The mare is restless, breaks into a sweat, urinates frequently, and lies down and gets up. Sometimes a foal may be dropped without warning.

ARRANGING A FOALING PLACE

Arrange a place for the foaling 7 to 10 days before the event. During mild weather, a clean open pasture where the ground is dry and warm is ideal, as there is little danger of infection or mechanical injury, and there is no other livestock.

When weather or space does not permit ideal conditions, consider a foaling stall. This should be at least 12 feet square and have a smooth, well-packed clay floor. It should be free of obstructions, such as low mangers or hay racks, on which the mare and foal might be injured. If possible, locate it away from other occupied stalls.

Scrub the foaling stall, the manger, and grain boxes with boiling lye water (8 ounces of lye to 20 gallons of water). They should be thoroughly dry before use.

Sprinkle the floor lightly with air-slacked lime. A clay floor may be harder to keep smooth and clean than one made of concrete, but a clay floor is much better for the foal's hooves and provides safer footing for both mare and foal.

See that the foaling stall has plenty of clean, fresh bedding. Stable the mare in the foaling stall at night for 7 to 10 days before the foal is expected, to accustom her to the new surroundings.

FEED AT FOALING TIME

Reduce the mare's grain allowance slightly just before the foal is due. Use light and laxative feeds liberally, especially wheat bran. If the mare seems constipated, feed her a wet bran mash.

Mares that have been properly fed and exercised, and have foaled before, usually experience no difficulty in delivering. Young mares foaling for the first time, old mares, and mares that are either overweight or thin and rundown may have difficulty. A good attendant will be near the mare at foaling time but not where she can see him. This is because some mares seem to resent the presence of others and will delay foaling as long as possible. An attendant helps to prevent injury to the mare and foal, and can call a veterinarian if necessary.

PARTURITION

Foaling begins with the rupture of the outer fetal membrane, followed by the escape of a large amount of fluid. The inner membrane surrounding the foal appears next, and labor becomes more marked.

A mare foals rapidly if the presentation of her young is normal. The entire birth process generally does not take more than 15 to 30 minutes. The mare will usually be down at the height of labor. Most often the foal is born while the mare is lying on her side with her legs stretched out.

In normal presentation of the foal, the front feet with heels down appear first, followed by the nose and head, then shoulders, middle (with back up), hips, and finally hind legs and feet.

Summon a veterinarian at once when the presentation is not normal. There is great danger that the foal will smother if birth is delayed.

The mare should expel the afterbirth 1 to 6 hours after foaling. If it is retained longer, or if the mare seems lame, blanket her and call a veterinarian.

Remove the afterbirth from the stall immediately and burn it or bury it in lime.

FEED AND WATER AFTER FOALING

The mare may be hot and feverish immediately upon foaling. Give her small quantities of lukewarm water at intervals. Do not allow her to drink too much.

Feed the mare lightly with laxative feeds for the first few days. Her first feed can be half a ration of wet bran mash with a few oats or a little oatmeal soaked in warm water. A combination of bran and oats makes the best grain ration for the first week. The milk flow, the demands of the foal, and the appetite and condition of the mare should govern the amount you give her.

Usually the mare can be back on full feed within 7 to 10 days of foaling.

DRYING UP THE MARE

Foals are luckier than calves. They are kept with their mother far longer and are reared as nature intended, on mother's milk. Most are weaned between 4 to 6 months of age, however, and this should be done when the moon is in Sagittarius, Capricorn, Aquarius, or Pisces. Allow the foal to nurse in a fruitful period for the last time, when moon-Venus aspects are operating.

To wean a foal successfully, it is important to do so when the moon is in a sign of the zodiac that does not rule vital organs. These good signs are Sagittarius, Capricorn, Aquarius, and Pisces, which rule the thighs, knees, ankles, and feet.

Cut the ration of the dam in half a few days before the separation. This helps her udder to dry up without difficulty.

Rub an oil preparation (camphorated oil or a mixture of lard and spirits of camphor) on the udder. Take the mare from the foal and place her on grass hay or less lush pasture.

Examine the udder at intervals and place oil on it, but do not milk it out for five to seven days. The udder will fill up and get tight, but do not milk it out. At the end of five to seven days, when the bag is soft and flabby, milk out the little secretion that remains (possibly half a cup).

If, by using creep or a separate grain box, the foal has become accustomed to eating a considerable amount of grain and hay, weaning will cause only a slight disturbance or setback.

Move the mare to new quarters, away from the stall it shared with the foal. Remove anything in the stall on which the foal might hurt itself during the first unhappy days that it lives alone. Make the separation of foal from mare complete and final. If the foal sees, hears, or smells its dam again, you will have to begin the separation process all over again.

Turn the foal out on pasture after a day or two. Do not run weanlings with older horses.

BREEDING AFTER FOALING

Mares usually come into heat 7 to 11 days after foaling, although the range is from 3 to 13 days. Some horsemen plan to rebreed mares during this first recurrence of heat on about the ninth day if foaling has been normal and there has been no discharge or evidence of infection.

It is believed that mares rebred at this time are more likely to conceive than if bred later. But, an increasing number of good horsemen prefer to rebreed during the heat period after foaling heat, between the 25th and 30th day of foaling.

THE FOAL

Immediately after the foal has arrived and has started breathing, rub it thoroughly and dry it with warm towels. Then place it in a corner of the stall on fresh, clean straw. If this corner is in the direction of the mare's head, she will probably be restless.

Be sure to protect the eyes of the newborn foal from bright light.

If left alone, the navel cord of the newborn foal generally breaks within 2 to 4 inches of the belly. If not, cut it about 2 inches from the belly with clean, dull shears or scrape it in two with a knife. A torn or broken blood vessel will bleed but little; one cut directly across may bleed excessively. Immediately treat the severed cord with tincture of iodine (or other reliable antiseptic), then leave the mare and foal alone so they can rest and gain strength.

FIRST NURSING

A strong, healthy foal is on its feet and ready to nurse usually within ½ to 2 hours after birth. Take advantage of this time to wash the mare's udder with a mild disinfectant, then rinse thoroughly with clean, warm water.

A big, awkward foal may need assistance and guidance when it nurses for the first time. However, if the foal is stubborn, forced feeding will be useless. Back the mare onto additional bedding in one corner of the stall and coax the foal to the teats with a bottle and nipple. An attendant should hold the bottle while standing on the opposite side of the mare from the foal.

Give the mare's first milk to a very weak foal even if it is necessary to draw this milk into a bottle and to feed the foal a time or two by nipple. Sometimes an attendant must steady a foal before it will nurse.

For the first few days following parturition, the dam secretes colostrum milk. This differs from ordinary milk in that it is more concentrated, higher in protein content (especially in globulins), and richer in vitamin A.

Colostrum is different from ordinary milk in two other ways: It contains antibodies that protect the foal temporarily against certain infections, and it is a natural purgative that removes fecal matter accumulated in the digestive tract.

Never dissipate these benefits of colostrum by "milking out" a mare shortly before foaling time.

BOWEL MOVEMENT

Regulation of the bowels of the foal is very important. Both constipation and diarrhea (scour) are common ailments.

Excrement impacted in the bowels during prenatal development can kill the foal if not eliminated promptly. A good feed of colostrum usually induces natural elimination. However, especially when foals are from stall-fed mares, this is not always the case.

Diarrhea (or scour) in foals results from either infectious diseases or dirty surroundings. It is caused by an irritant in the digestive tract that should be removed. Give an astringent only in exceptional cases and only on the advice of a veterinarian. Honey, formed into thick balls, is considered helpful against diarrhea. (You can mix honey with slippery elm powder.)

Conditions that cause diarrhea are contaminated udder or teats, nonremoval of fecal matter from the digestive tract, fretfulness, temperature above normal in the mare, an excess of feed affecting the quality of the mare's milk, a cold damp bed, and continued exposure to cold rains. If the foal is scouring, reduce the mare's ration and take away part of her milk at intervals by milking her out.

CARING FOR THE SUCKLING FOAL

Weather conditions permitting, there is no finer place for a mare and foal than on clean pasture. As with all young animals, milk from the dam assures the foal the best possible start in life.

When the foal is between 10 days and 3 weeks old, it begins to nibble on a little grain and hay. To promote thrift and early development, and to avoid any setback at weaning time, encourage the foal to eat supplementary feed as soon as possible.

The foal should be provided a low-built grain box especially for this purpose; or, if on pasture, the foal may be creep fed.

Rolled oats and wheat bran, to which a little honey or brown sugar has been added, is especially palatable as a starting ration. It is a practice among the Arabs to give honey to their best purebred Arabian foals, about 2 tablespoons apiece. They consider that honey not only is mildly laxative, body cleansing, soothing, and restorative but also gives strength and stamina. In older animals, it is thought to enhance fertility as well.

A mare and foal at pasture

Provide crushed or ground oats, cracked or ground corn, wheat and bran, and a little linseed meal (oats, 30 pounds; barley, 30 pounds; wheat bran, 30 pounds; linseed meal, 10 pounds) later. If none of these is easily obtainable, feed a good commercial ration.

Give the foal good hay (preferably a legume) or pasture, in addition to its grain ration. A normal, healthy foal should consume about ½ pound of grain daily per 100 pounds of live weight at 4 to 5 weeks of age. This ration should be increased about ¾ pound or more per 100 pounds of live weight by weaning time.

Foals should grow to half their mature weight during the first year. Most breeders of Thoroughbreds and standardbreds plan to have their 2-year-old animals at full height. Such results require liberal feeding from the beginning. Good growth must be made at this time; otherwise, it won't be achieved in the animal's lifetime.

AFTER WEANING

Unless the mare is sickly, the foal should not be weaned before at least 4 months of age. Six months is better, and if the mare is healthy and has not been served again immediately following the birth of the foal, feeding even into the ninth month is beneficial. Of course, the milk the foal takes after the fourth month is a supplement to other foods it will be receiving.

Dandelions in the pasture are considered beneficial, as is dried nettle hay, which you can chop and add to their rations two or three times a week. Fresh nettles, chopped, are important for red blood and strong nerves. Carrots cleanse the body of worms. Marigold flowers, hop shoots, and strawberry foliage act as a tonic. (They're also good for adult animals.)

The winter diet should be silage, hay, and sliced or pulped roots, along with a moderate amount of crushed wheat and barley. Mix this with a small amount of chaff for roughage.

A daily gallon or so of a mash consisting of bran and molasses with a little linseed meal, to which you can add sliced carrots, rutabagas, beech mast, hawthorn haws or crushed rose hips, seaweed powder, watercress, and a few aromatic seeds such as anise, fennel, and dill, is excellent. Give ripe apples and pears in small amounts. Mulberry leaves and carrots will cleanse both foals and adult animals of worms.

CASTRATION

A veterinarian should perform this operation. Select a date when the moon is not in Virgo, Libra, Scorpio, or Sagittarius. The procedure is best done when the date is within 1 week before or after the new moon.

Most horsemen prefer that the operation be performed when the animal is about a year old, although a colt may be castrated when it is only a few days old. While there is less danger to the animal and much less setback with early altering, it often results in imperfect development of the foreparts. Leaving the colt entire for a time yields more muscular, bold features and better carriage of the foreparts.

Weather and management conditions permitting, the time of altering should be determined by the development of the individual. Underdeveloped colts may be left entire six months or even a year longer than overdeveloped ones.

TRAINING THE FOAL

For whatever purpose the horse is intended, his training must begin early. A foal will not need "breaking" if it has been trained properly, and it will be a better disciplined, more serviceable horse. Give the foal its lessons one at a time and in proper sequence, and be sure your pupil masters each lesson.

Put a well-fitted halter on the foal when it is 10 or 14 days old. When the animal is accustomed to the halter, after a day or so, tie the foal securely in the stall beside the mare. Try to keep the foal from freeing itself from the rope or from becoming tangled up in it. Leave the foal tied 30 to 60 minutes each day for two or three days. Groom the animal carefully when it is tied. Rub each leg and handle each foot so that the foal becomes accustomed to having its feet picked up.

After the foal has been groomed, lead it around with the mare for a few days, then lead it by itself. Lead it at both the walk and the trot. Many breeders of Thoroughbreds teach a foal to lead simply by leading it with the mare from the stall to the paddock and back again.

At this stage of training, be sure the foal executes your commands to stop and go as soon as you give them. When halted, make the foal stand in show position, squarely on all four legs with its head up.

Be patient, gentle, and firm in training the foal. There will be times when the foal is stubborn; never let your temper get the better of you. Do not keep the foal working for too long — short training sessions are best.

The day will come when you can saddle your young horse with ease because you have followed a good training program. Saddling and harnessing will just be additional steps. A good time to harness and work the horse for the first time is during the winter as a rising 2-year-old.

HEALTH PROBLEMS

Worms. These are a common affliction in the domestic horse. The four principal kinds are thread, round, tape, and bot. Red threadworm is the most common and the most troublesome. Symptoms are an unthrifty appearance with a tight coat and general emaciation despite a greedy appetite.

For mild cases of worms in horses and foals, garlic treatment given morning and night is usually sufficient. Just add 3 or 4 grated roots of garlic (bulbs) to a mash of bran and molasses, and continue this feed for several weeks. All worm treatments are best given when the moon is waxing and near full. (The worms are then breeding and stirring, and easier to dislodge.) Another good idea is to feed plenty of pulped carrots, first removing any stringy ends. Further, provide access to couch grass and brambles.

If worm infestation is heavy, fast the horse for two days, then night and morning give a dosage of balls made from 3 or 4 grated garlic bulbs bound with flour and honey. Then feed a warm mash of bran with more molasses and turn the horse out to grass. Feed also seedy fare such as pumpkin, mustard, seeded parsley, and melons.

Lice. Internal treatment will be necessary for a cure. The bloodstream of the horse should be well saturated with garlic, given daily in strong doses. Externally, scrub the animal all over with soap and water to which pine disinfectant and some soda have been added. Then rub in a strong brew of quassia wood chips. When the animal is dry, dust well with derris powder.

The Arabs use oil of eucalyptus mixed into a little ammonia: 2 teaspoonfuls of eucalyptus oil and 1 tablespoonful of ammonia to 1 quart of warm water. Rub this into the body hair, carefully avoiding the eyes. This will usually clear serious lice infestation.

Ringworm. "Ringworm" is the general name for several skin diseases caused not by worms but by tiny plants, or fungi. Take tobacco ashes, wet with vinegar, and apply to the afflicted area.

CHAPTER 9

TAURUS, THE BULL

The 12th house rules great cattle. This includes beasts of burden such as camels and elephants.

In all probability, modern European cattle descended from two different types of wild cattle, both of which belonged to the genus *Bos*, species *taurus*.

Humans have raised cattle for thousands of years. Cattle raisers once followed their herds as the cattle searched for fresh pastures. Some of these herdsmen later settled in one place and began feeding their cattle grain, which they raised, in addition to grass. In many agricultural societies, a man's wealth was appraised by the herds he owned.

Here in America, the word *cattle* usually means cows, bulls, steers, heifers, and calves. The cow is the female and the bull is the male. Steers are males whose reproductive organs have been removed by castration. A young cow is called a heifer until she gives birth to a calf. A calf may be a young cow or a young bull. A group of cattle is called a herd.

Both beef and dairy cattle that can be traced through all their ancestors to the original animals of a breed are called purebred. A registered animal is one whose family history has been recorded with the association for that particular breed.

BREEDS OF BEEF CATTLE

The principal breeds of beef cattle are the Hereford, Shorthorn, Aberdeen-Angus, Brahman, or Zebux, and the Santa Gertrudis. A new type of beef cattle called the Beefalo has recently been developed.

The Santa Gertrudis was the first distinct breed of cattle produced in the United States. In the 1920s and 1930s, the King Ranch at Kingsville, Texas, crossed Brahmans and Shorthorns to develop this breed.

Shorthorns are a dual-purpose cattle and include three strains. The term Shorthorn alone is applied to cattle raised for meat. Milking Shorthorns are raised for both beef and milk, Polled Shorthorns (which are hornless) are raised for beef.

For more than a hundred years, cattlemen experimented with buffalo crosses but with little success. Then the Beefalo was developed by D. C. Basolo of Tracy, California, who took more than eight years to come up with the right combination. The Beefalo hybrid is three-eighths buffalo, three-eighths Charolais (a breed of white beef cattle of French origin introduced into the United States in 1936), and one-fourth Hereford.

The Beefalo gains weight faster than others and can finish out on grass rather than grain. Even when entirely fed on grass, Beefalo compares favorably with other breeds as to taste.

For those concerned with their cholesterol count, Beefalo is good news again, higher in protein and lower in fat than other beef.

Beefalo is a distinct breed, and it takes five breedings to get pure Beefalo stock. But ranchers do not need pure Beefalo stock. The calves that gain weight fastest are from the first cross, that of a Beefalo bull and a cow of another breed, and the cow can be of *any breed.*

Actually, the calves look not very different from regular beef for Beefalo, even with their three-eighths buffalo blood; they more resemble the domesticated kind than they do the wild buffalo that contributes to their ancestry. They are relatively small at birth, weighing just 40 to 70 pounds.

Do you see how significant this is?

Beefalo

The homesteader who keeps a dairy cow for milk can obtain semen, have a veterinarian perform the artificial insemination, and raise out a good-quality beef animal to fill the freezer with steaks, roasts, ground meat, and all the rest of the goodies we so enjoy. And this can be done on a day of the homesteader's choosing so that the calf will be born in one of the fruitful periods of Cancer, Scorpio, or Pisces to ensure favorable growth

DAIRY COWS

Dairy cattle include the Holstein-Friesian, which farmers favor because it produces more milk than other breeds, though lower in butterfat.

Jersey cattle are the smallest breed, ideal for the small homestead or the small family. Jerseys produce less milk than the other major breeds, but their milk contains the most butterfat. You can expect a thick mass of cream to rise to the top of a container of Jersey milk.

Guernseys are slightly larger than Jerseys. They produce a little more milk but rank second to the Jerseys in butterfat content. Ayrshire milk production and butterfat content rank between Holsteins and Guernseys.

Another breed is the Brown Swiss, which are larger than most dairy cattle. Brown Swiss produce fairly large amounts of milk that is pure white and rich in nonfat solids, minerals, and especially lactose, or milk sugar. These qualities make the milk excellent for cheese.

ABOUT MILK

Milk is ruled by the moon, Venus, the sign of Cancer, and the 4th house. The moon rules the basic fluid of milk. Jupiter adds the fatty constituents; the Sun adds casein, the clotted protein of milk; and Saturn adds the salt. The period of lactation is ruled by the sign of Cancer. Venus rules bulls and calves.

Milk contains proteins, carbohydrates, fats, minerals, and vitamins — all the nutrients necessary for healthy bodies and sturdy teeth and bones. Other foods contain these same nutrients but only milk has them in amounts such that they can work together as a team.

Milk is especially important to those born under the signs of Gemini, Virgo, Libra, Scorpio, Aquarius, and Pisces. Pisceans should use it liberally.

BUYING A DAIRY COW

Buy a cow on a day when the moon is not aspected with Saturn. You need not set your sights on a fine, registered, purebred animal. A well-cared-for grade should give quite satisfactory milk production. Your decision quite likely will be influenced by what is available. She may give as little as 2 gallons or as much as 7 or 8, but you will find that rich, sweet, fresh milk very worthwhile.

It's a good idea to get a veterinarian's certificate, guaranteeing her to be free of an infectious disease such as tuberculosis or Bang's. Check to see that her teats are neither too large nor too short; either condition makes for difficult milking. Also have her checked for mastitis.

Temperament is another important factor. Check her gentleness by moving around the cow. You may even wish to milk her before buying.

Naturally, it is much more expensive to keep a cow if you must purchase all her feed. I don't advise keeping one if you must do this. But if you have a couple of acres of good pasture, you'll find that is plenty for her, and your vegetable garden and orchard will provide much that she consumes during the summer months. You may also be able to get some free lettuce or cabbage trimmings from a grocery store.

In figuring the cost of a cow and her keep, consider what she will produce. Of course, there is first of all the milk, the luxury of having real cream, and perhaps making cheese. You may even have some surplus to sell. She will also produce a ton or so of manure, mixed with used bedding, to the great benefit of the compost heap and eventually the garden. Cow manure, like that of hogs, is relatively wet and correspondingly low in nitrogen but because of its high water content and low amount of nitrogen, it ferments slowly. It is commonly regarded as a *cold manure.*

GETTING READY

First, do you have the space for pasture? If you do, it should be well fenced; otherwise, you may wake up some morning and find your cow in the vegetable garden. She must also have a stable.

MILKING

The milking procedure is not difficult to learn. It is largely a matter of muscle development in the hands and arms, and that's best acquired by practicing on the job. A cow is usually milked from her right side, but a gentle animal can get accustomed to a milker on either or both sides. Four hands are quicker than two.

A cow, like a goat, should be milked regularly, but you may suit the hours to your convenience, within reason. And it does not have to be early; 8 or even 9 o'clock is fine. She can again be milked in the evening after supper.

Cleanliness about the milk is crucial. Keep the cow's flank and udder clipped and clean, so particles of dirt or bedding will not drop into the bucket.

Milk regularly.

For best flavor the milk should be quickly cooled. Place it in crocks or wide-mouthed jars that will be easy to skim. The cream may be refrigerated and whipped right in the jar.

You can usually milk in 10 to 20 minutes. This time includes feeding, washing the udder, and the actual milking process. It takes just a few minutes more to strain the milk through a filter, put it away in the refrigerator to cool, wash the bucket, and put it out for the disinfectant action of the sun.

As the owner of even one cow you can enjoy another luxury, butter. You may wish to invest in a small churn, but butter can also be made by whipping with an eggbeater, in a blender, or even just by shaking a covered jar until the butter comes. You then rinse the butter and beat out the buttermilk (wonderful for a cold drink on a hot summer day). Add a little sea salt if you like. Some people prefer to let the cream sour before making butter, and you may wish to try this. A sweeter butter results when the cream is used without being permitted to sour.

BREEDING

Your cow reaches her peak of production about a month after she freshens (calves). Choosing a good sign, it is best to breed her again about two to four months after calving. Plan to have her bring her calf in Cancer, Scorpio, or Pisces; Taurus or Capricorn is a second choice. The gestation period in cattle is almost exactly the same as it is for a human being, 283 days.

There is a great advantage in artificial insemination, particularly if you are keeping only a few milk cows. First of all, you don't have the cost of a bull and his upkeep; second, you have a wider choice of breeds.

For the first calf of a Jersey heifer, for example, insemination of either Angus or Brangus semen is advisable because the calves are usually small enough to be delivered without problem. Never use semen of one of the big breeds for a small cow. In the Brahman and Brangus breeds, there is the added advantage of having the calves' heads narrow at birth. If you want to raise out your calf for beef rather than milk, consider the Beefalo breed.

DRYING OFF

Milk production decreases slightly after your cow is bred because from then on the unborn calf begins to take more and more of her nutrition. Four to eight weeks before the calf is due, dry up the cow by decreasing the feed and discontinuing the milking.

If your cow is still producing copiously, start by milking only once a day for a week and cut her grain ration in half. Then skip one day's ration and milking; milk out only enough to relieve her udder a bit if it is very tight. Next, do not give any grain at all for a week but let her graze and eat some hay. Herbs such as asparagus, cranesbill, periwinkle, and mint (particularly water mint) aid in decreasing milk flow.

After your cow is dried off, give her a half-ration of grain each day until a week before the calf is due. At this time, increase her feed to about 10 pounds. Within a week or so of calving, she can be put back on full rations.

CALVING

Calving is a natural process with cows, and nature has equipped them to deliver their young with little assistance or trouble, except in rare instances. The best help you can give is to provide your cow with a clean, dry, warm stall and plenty of clean bedding.

In a normal birth, the front feet appear first. The calf's nose appears just about the same time as the knees. After the head and front shoulders are out, birth is usually over in a few minutes. However, if she is in labor for several hours without delivering, she may need help. Occasionally a calf is born back feet first, or the calf's head may be bent back (thus preventing birth), or it may be in some other abnormal position.

Should the back feet appear first with heels up, your cow likely will need help. Get a veterinarian if possible. If you can't get help, have a light rope handy. Loop this over the calf's feet and pull *downward* (never up), toward the cow's feet. After the rear end of the calf is out, pull the animal all the way out quickly so it won't be smothered. After you have wiped the membrane and mucus from the calf's nose, let the cow take over the rest of the clean up.

CARING FOR THE CALF

A few hours after the calf is born, treat its navel with iodine and determine whether the calf has nursed. It may need help to get started; sometimes the teats are too large for the little calf's mouth. You may even need to milk the cow and give the calf its first feeding from a bottle.

I consider it imperative that the baby calf be fed the colostrum milk for the first week or 10 days. This milk is not good for human consumption but is needed for the calf's well-being. If there is more colostrum than the calf can use (and with a good milker there usually is), give it to any other calves or dilute it with water and feed it to other animals such as pigs and even dogs and cats. A new calf seldom needs more than a half gallon at a feeding, or a gallon a day. Sometimes feeding regulated amounts from a nipple bucket is best.

When the calf is about a week old, introduce it to the water pail, a little grain feed, and some fresh hay. You'll be surprised how quickly it will get the hang of it.

SHOULD THE CALF NURSE?

There are two schools of thought on this matter. Many feel that suckling of the cow is beneficial to both the cow and the calf; the shrinkage of the uterus after parturition is dependent on the nerve stimulus provided by the sucking calf, while the calf develops stronger jaws, stronger legs, and a better digestive tract.

On the other hand, allowing the calf to suckle may cause some problems. Whole milk from Jersey cows is so rich that a young calf can get scour. It is sometimes better to give the calf a bottle of skim milk containing only a small amount of cream, rather than whole milk or even a commercial milk substitute.

Consider, too, the condition of the cow. Today's modern, dairy-type cows have been bred and upgraded to produce more and more milk. Because of this, their udders are apt to be sensitive and thin-skinned (unlike the udders of the beef breeds). In sucking, the calf instinctively bumps the udder, and too much of this can cause bruising and eventually bring on mastitis problems.

Another good reason for separating the cow from the calf is that the cow will come back into heat and you can rebreed her sooner. Usually her first heat occurs three or four weeks after her calf is born, each heat being about 21 days apart. If you have two cows, stagger the breeding dates so that you will have one cow in production at all times.

SURPLUS MILK

Undoubtedly there will be times when you have more milk than you can use. Selling part of it will help pay for feed or, in an overall farm operation, you can use it to feed other animals. You can grow out two or three pigs, at least partly; and chickens thrive on skim or surplus milk. You may even, in time, learn how to make various cheeses, the best way of all to use it up. And then, of course, there is ice cream!

A CALF FOR YOUR FREEZER

When you breed your cow to a beef-type bull and have a spring-born calf, it will be ready for the freezer by fall.

Veal is the flesh of calves from 2 to 14 weeks old. Calves that are older are usually slaughtered (or sold) as yearling beef. Veal is more tender than beef and contains a higher percentage of water and less fat.

Beef is the flesh of older cattle. Some prefer it to veal because of its delicious flavor. Good beef can be distinguished from poor beef because its fat is white and the lean meat has a bright, cherry red color.

MEAT AND YOU

Meat is an excellent source of a number of vitamins. Nearly all the B-complex group is found in beef. The most important of these are thiamine (B_1), necessary for growth and the proper function of the heart and nerves, and riboflavin

(B_2), important for healthy skin and properly functioning eyes. Nicotinic acid, important for the prevention and treatment of pellagra, is present in meat, and liver is a source of vision-aiding vitamin A and vitamin D, which aids in building bones and teeth.

Meat also contains certain minerals used by the body. Iron and copper are needed in the hemoglobin of red blood corpuscles and are found in nearly all meat, but are especially abundant in liver. Most meats also contain phosphorus, a mineral useful in building strong bones and teeth.

Meat is necessary to you in greater or lesser amounts according to your birth sign, and for some few it's better to avoid meat altogether.

Aries	March 21–April 19	Include little meat in diet.
Taurus	April 20–May 20	Eat meat sparingly. (Taureans like meat but tend to overeat.)
Gemini	May 21–June 20	Eat meat once a day.
Cancer	June 21–July 22	Eat meat in moderate amounts.
Leo	July 23–August 22	Avoid meat. Eat fruits, vegetables, cereals.
Virgo	August 23–September 22	Eat meat sparingly.
Libra	September 23–October 22	Eat game, poultry, and fish.
Scorpio	October 23–November 21	Avoid meat. Eat fruits, vegetables, cereals.
Sagittarius	November 22–December 21	Eat meat in moderate amounts.
Capricorn	December 22–January 19	Meat and other heat-producing foods are important.
Aquarius	January 20–February 18	Eat meat once a day.
Pisces	February 19–March 20	Eat meat once a day in moderate amounts.

CATTLE AND COSMOLOGY

Animals, bulls in particular, are easiest to handle when the moon is in Taurus, Cancer, Libra, or Pisces. Avoid the full moon. Buy during the first quarter. If an animal is to be castrated, select a date when the moon is not in Virgo, Libra, Scorpio, or Sagittarius. Again, avoid the full moon.

CHAPTER 10

THE SIGN
FOR SWINE

Swine, being small animals, are ruled by the sixth house,

which is affiliated with the sixth sign, Virgo.

In China, the domestication of the pig began as long ago as the Neolithic era. In fact, there is a legend that the first pig ever to be cooked, or roasted, was in China.

Wild hogs roamed throughout Europe and some other parts of the world as far back as 6 million years ago. It is believed that man began taming and raising hogs descended from these animals during the Stone Age, about 7,000 years ago, and there are good reasons for homesteaders to do the same today.

Pig-keeping is less of a chore than you may think. A pig is one of the easiest animals to keep and feed. He is a great scavenger and a fine natural garbage can. Pigs are smart animals, ranking in intelligence even above the horse, and, if properly cared for, they are the cleanest of barnyard inhabitants.

Pigs also contribute generously to that essential of the compost heap, manure. Each ton of manure contains 500 pounds of organic matter, 10 pounds of nitrogen, 10 pounds of potassium, and 5 pounds of phosphorus. Just dwell on that idea for a moment, and think how it will enrich your soil.

BREEDS

Farmers in the United States raise some 23 breeds of hogs. Hogs are roughly described as meat-type, bacon-type, or fat-type, but all are raised primarily for meat.

The most widely raised breeds of hogs are the Duroc (meat-type), Poland China (meat-type), Hampshire (meat-type), American Landrace (bacon-type), and Yorkshire (bacon-type). Chester Whites are fine meat producers.

No one breed is greatly superior to another in its ability to produce meat, grow rapidly, or to bring large litters. A herd of good meat hogs can be developed from any of the above breeds.

MEET THE PIG

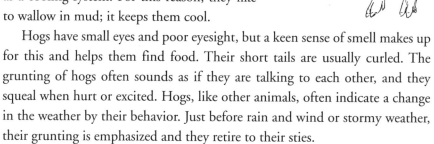

The hog's body is stout and heavy and covered with coarse, bristly hair. The head and short, thick neck extend forward in a straight line from the body. The head ends in a snout. The thick skin of a hog has no sweat glands to act as a cooling system. For this reason, they like to wallow in mud; it keeps them cool.

Hogs have small eyes and poor eyesight, but a keen sense of smell makes up for this and helps them find food. Their short tails are usually curled. The grunting of hogs often sounds as if they are talking to each other, and they squeal when hurt or excited. Hogs, like other animals, often indicate a change in the weather by their behavior. Just before rain and wind or stormy weather, their grunting is emphasized and they retire to their sties.

At birth, piglets usually weigh in at about 2½ pounds and generally double their weight the first week. Fully mature boars may weigh more than 1,000 pounds and sows around 900 pounds. However, most hogs are slaughtered or marketed when they are 5½ to 7 months old and weigh from 180 to 240 pounds.

The broad, leathery pad of the hog's snout includes the nostrils and is very sensitive to the touch. If at any time a hog should prove too aggressive, this is where you should place a light blow to turn the animal.

The snout is often used by the hog to root (dig for vegetable roots) for some of their favorite foods. In France farmers use specially trained hogs to search out truffles, a type of subterranean fungus considered a great delicacy.

Hogs cause damage to pastures when they root. For this reason, the farmer sometimes slips a ring through the nostrils. This causes pain if the hog roots. Hogs, if permitted, will even root up young trees. On the other hand, the judicious use of rooting hogs in an older orchard can be beneficial. The rings are removed and they are allowed in during the winter season. The area is not only thoroughly dug but also enriched with their droppings. In the spring the pigs are removed and the ground is leveled. Grass quickly grows again in the fertile soil, and the quality of the fruit is greatly improved both from the aeration (digging) and from the manure.

Hogs have canine teeth, called eyeteeth, that often develop into sharp tusks, particularly in the mature male. These teeth serve as both tools for digging and weapons for fighting. Because they can cause serious injury, farmers usually cut off the tusks with special clippers. Dental work of this type is best done when both the sun and the moon are in barren signs and the moon is decreasing, fourth quarter preferred. Barren signs are Leo, Aries, Virgo, Aquarius, Gemini, and Sagittarius, in order of their degree of nonproductive qualities.

Hogs are not particularly aggressive and prefer to protect themselves by running away, but they can be dangerous when cornered. Take care with a large boar or a nervous sow with a litter.

STARTING A HERD

Begin with a couple of shoats (young, freshly weaned pigs) in the spring. If you don't really want to do any breeding yourself, it might be simpler just to repeat this each spring, buying from a local breeder and fattening your pigs for fall hams. Check the classified ads in your local paper for leads, or subscribe to a regional farm magazine, or ask your county agent.

Choose pigs about 8 weeks old, selecting the largest and huskiest in the litter. If you buy barrows (castrated males), pick those with short legs, compact shoulders, and plump hams. Animals should weigh between 25 and 40 pounds. Buy when the moon is not aspected with Venus.

On the other hand, if you breed your own, the next year's stock will cost only your labor in growing feed, and you may also have some pigs to sell. As with other livestock, start small (two the first year and maybe four the second) and get to know and understand your animals and their requirements. Butcher three and keep one sow to breed. By that time you'll have some idea of how to house, feed, and handle pigs. You'll know whether you want more of the same or just the few you've been keeping for family use.

HOUSING

A hog pen with a raised, slatted floor is helpful if you must keep your pig in a small space. If possible, place the pen on a gentle slope for good drainage. For the floor, use two-by sizes, spaced ⅞ inch apart, which rest on concrete blocks. Make sure it's high enough that manure may easily be raked out from underneath to avoid fly and odor problems. Sprinkle lime on the ground occasionally.

SWINE NUTRITION

Of all farm animals (with the exception of chickens), the pig is the most efficient converter of feed to meat, but you'll still need to feed well-balanced ratios and prevent wastage.

Many a commercial piggery feeds entirely on garden wastes or clean kitchen or restaurant garbage. You may be able to make a connection with a small restaurant and cut down on feeding costs.

If you have the space, pigs do exceptionally well on pasture, which can supply 20 to 30 percent of their total requirements. The best pasture consists of alfalfa, clover, and rape, or a mixture of rape and oats.

Good summer pastures can be soybeans, sorghum, or Sudan grass; later in the season, ryegrass and winter barley are good. If your pastures are plagued with quack grass, you can depend on pigs to root it out, roots and all, leaving your land in condition to plant a better crop.

An acre of good forage is sufficient for 20 pigs with an average weight of 100 pounds. Construct a movable house to provide shelter and shade. A small, clean shelter open at one side is usually adequate. (See Kelly Klober's *A Guide to Raising Pigs* for housing recommendations.)

Pigs that are allowed to root naturally are healthiest. Perhaps they are getting needed trace minerals that are contained in the earth. (Soil may contain other unknown substances that are vital to the assimilation of nutrients.)

If pigs are confined, give them a mineral mixture of 2 pounds of oyster shell or limestone flour to 2 pounds of bonemeal and 1 pound of salt at the rate of 1 pound for every 5 pounds of feed.

ENERGY FROM CEREAL GRAINS

Most of the energy in a swine ration is supplied by a cereal grain. Grains most commonly used are corn, milo, barley, and wheat.

All cereal grains are too low in protein quality to meet the requirements of swine. Therefore, supplements are a good idea. Some of these are soybean meal, fish meal, and meat and bone scraps or tankage (by-products of the meat packing industry).

MINERALS

There are 13 minerals required by swine. Eight of these are not present in large enough quantity in natural feedstuffs to meet their needs. They are calcium, phosphorus, sodium, chlorine, copper, iron, iodine, and zinc. Trace mineral mixes are commonly used to provide these, as is mineralized salt.

LAST BUT NOT LEAST, WATER

A plentiful source of clean water is actually more important than any other nutrient. Supply water in sufficient amounts at all times for maximum performance. The water requirement for swine is about 1 gallon per hundredweight daily or twice as many pounds of water as feed. Always provide more during hot weather.

Hogs tend to gain weight more quickly and more economically when they have plenty of fresh water during freezing weather.

REPRODUCTION

You can mate your hogs when they are about 8 months old. Litters vary from 5 to 10, but may be as many as 27 or so. Hogs may be bred twice a year, in both spring and fall, but for the beginner I recommend once-a-year spring breeding. Borrowing or renting a boar for the occasion is much more economical than trying to keep your own when only a sow or two is kept.

Choose a boar with a record of meat-producing characteristics, in addition to a history of siring large litters. The pedigree of the boar influences more pigs than will any single sow.

Do not underrate the sow, though, for she should be chosen for the same traits as the boar. In addition, she should have no fewer than 8 teats, preferably 12. This characteristic is heritable, as is her ability to produce milk in large quantities.

Sows come into heat on an average cycle of 21 days. Older sows remain in heat longer than younger ones; the usual time is two or three days. Ovulation usually occurs the second day of estrus. The sow will be restless and grunt loudly. To ensure fertilization, mate her twice. Allowing a 24-hour break in between. Feed the sow well during gestation. This means good pasture, if possible.

Swine have a short breeding cycle, 112 days, and the average litter is 5 to 10. Breed so that the litter is born in any one of the fruitful signs of Cancer, Scorpio, or Pisces (water signs), with Taurus or Capricorn (earth signs) as a second choice.

FARROWING

Several days before a sow farrows, wash it and move it to a clean pen. (*Note:* When you wash the sow, take particular care to remove the small plug of dirt that may be on the end of the teats.)

The sow is clumsy, unfortunately, and special care must be taken to keep her from crushing her young when she lies down. Some farmers use special farrowing stalls that confine each sow to a small space but still allow the pigs to move about freely. If the weather is very cold, install a heat lamp in a corner of the pen, or in a compartment called a pig brooder. This will attract the pigs when they are not feeding and keep them out of the sow's way. A sow rarely has difficulty farrowing, but if you are new to this and labor goes on for some time, call a vet or an experienced neighbor. As the pigs arrive, take the following steps:

1. Dry off each pig.
2. Remove the membrane from around the nose.
3. Tie the navel cord and disinfect with a germicide such as tincture of iodine.
4. Be sure the sow has cleansed, and burn or bury the afterbirth.
5. Help the pigs to nurse as soon as possible.
6. Do not let pigs chill. You may need to use a heat lamp.

FIRST DAYS AFTER FARROWING

You have some important chores to do on the first day after your sow has farrowed:

1. Clip the tips of needle teeth.
2. Ear-notch all pigs for identification.
3. Recheck temperature, ventilation, and heat lamps.
4. Be sure your sow has plenty of fresh water.
5. Even up litters, if necessary, if you have more than one sow. Transferring pigs to another litter should be done within a day or two.

The chores continue three days after farrowing and thereafter:

1. Prevent anemia by injecting the pigs with 100 mg of iron in the muscular tissue. The best place to jab the needle is in the neck area or under the foreleg.
2. Keep farrowing pens clean, bedded, and dry.
3. Toughen the pigs by gradually raising the heat lamps.
4. Gradually increase the sow ration. Self-feed at 7 days.
5. Castrate male pigs at 2 weeks of age, selecting a date when the moon is not in Virgo, Libra, Scorpio, or Sagittarius, and when it is within one week before or after the new moon.
6. Start creep feeding when pigs are approximately 2 weeks old.

> ## USING THE HEAT LAMP
>
> Keep baby pigs warm and dry. Optimum temperature for the baby pig for the first two or three days of life is 85 to 90°F. Reduce this temperature gradually, as the pig grows, to about 70 to 75°F.
>
> Heat lamps are helpful most of the year. A 250-watt heat lamp should be approximately 20 to 22 inches from the floor. If the heat lamp is placed near the sow, it should be at least 12 inches above her udder.

WEANING PIGS

1. Wean pigs when they weigh approximately 25 to 30 pounds. This should be done when the moon is in Sagittarius, Capricorn, Aquarius, or Pisces. Let the pigs nurse for the last time in a fruitful period.
2. Pen together pigs of the same size.

3. Pen pigs in small groups if possible. Twenty pigs per pen is best; do not exceed 40.
4. Provide warm, dry, and draft-free quarters.

REBREEDING SOWS

Usually sows come in heat 4 to 7 days after weaning the pigs. Rebreed the sows on the first heat period or pass them over to the next period, 18 to 24 days later, and rebreed them. After you have gained experience, you may wish to breed your sows more often than once a year.

After weaning, place the pigs where they can neither see nor hear the sow. The sow should be fed an increased amount of roughage to help dry up her udder.

FEEDING YOUNG PIGS

Sometimes, because of an unusually large litter, it may be necessary to bottle-feed some of the pigs. An old farmer's trick to wean the pig from the bottle is to put a cat or kitten with it and put milk in a pan. The pig will quickly learn to drink by watching the cat.

If you are raising only a few pigs, it is better to purchase starter rations or pigs. When they reach 30 to 75 pounds, let them join the rest of the herd in the pasture.

Wean a bottle-fed piglet by encouraging drinks with a cat.

SLAUGHTERING

While it is not difficult to slaughter chickens or rabbits, a pig is something else again. I suggest you make use of a custom butchering service that picks up the animal, slaughters it, processes and freezes the meat, smokes the hams and bacon, and renders the lard. If this is not possible, hire someone experienced to do this the first time and watch and learn. My husband usually asked a friend to assist him, whom Carl would help in turn. We cut up our porker on our dining table (covered with oilcloth) and package and freeze the meat ourselves, saving a considerable sum. In time you can do it this way, too.

Slaughtering is best done in late fall when the temperature is 32–35°F. Souring bacteria multiply rapidly at temperatures above 40°F. Slaughter the first three days after the full moon, preferably in the sign of Sagittarius, and meat will keep better, be tender, and have fine flavor. Avoid the sign of Leo.

The fourth quarter is the best time to salt and smoke pork. Meat that has been slaughtered during the increase of the moon, or in a moist sign (Cancer, Scorpio, or Pisces), is watery and will "spit" and curl up in the pan when fried. It will not keep well, either.

Hogs to be butchered should be taken off feed 24 hours before. Give them plenty of water and keep them from becoming excited.

PORK

Pork is a delicious food, rich in iron and B vitamins, especially thiamine, riboflavin, and niacin. It must always be cooked thoroughly. Trichinosis, a disease that comes from eating infected and undercooked pork, is caused by a tiny roundworm, the trichina, which enters the bodies of hogs through their food. Cook pork long enough at 137°F to kill any worms in the meat. The worms, if present, remain alive if the meat is undercooked, and enter the intestines of the person who eats it.

The U.S. government does not approve pork for various meat products unless the pork is stored 20 days at 5°F or lower or is quick-frozen to kill the trichina worms. If you do your own slaughtering, bear this in mind. Hogs get trichinosis chiefly by eating garbage that contains infected pork. Swine fed on grain or cooked garbage contract the disease much less often. The correct feeding of swine is the most important way to prevent trichinosis in humans.

FREEZING AND DEFROSTING

Cut large cuts of fresh pork (loin, shoulder, and leg) into convenient sizes for freezing. Package chops and steaks according to the number of servings needed. Shape ground pork into patties or package in portions for loaves or casseroles. Season ground pork after thawing, the flavor of most seasonings is intensified upon freezer storage. Wrap meat closely and seal tightly in moistureproof material, separating individual servings by a double layer of wrapping material. Label packages, noting date, cut, and weight (or number of servings).

Pork should be frozen quickly and stored at 0°F or lower. The recommended maximum storage period is 1 to 3 months for ground pork and 3 to 6 months for other fresh pork.

Cured and smoked pork such as ham, picnic, loin, shoulder roll (butts), and fresh pork sausage when frozen should not be kept in storage longer than 60 days. Bacon, bologna, frankfurters, canned hams, and canned picnics are not recommended for freezing.

CURING

Thoroughly chill fresh pork and process it within five days. Meat intended for curing should be kept cold but not frozen. Too much salt makes the meat dry and salty to the taste; not enough salt encourages spoilage. Therefore, weigh the meat and measure the curing ingredients. For every 100 pounds of dry-cured pork, trimmed hams, bacons, and shoulders, dry using 6–8 pounds of coarse-ground plain salt with 2 pounds of sugar and 2 ounces of saltpeter. It is very important to mix the ingredients thoroughly.

Divide this mixture into two parts; save one part for later, when the hams must be gone over again.

Begin by rubbing one part of the sugar-cure mix into all surfaces of the hams. Do this with a slow, kneading motion, taking care not to roughen the surface on the flesh side. Fit the salted meat into a clean crock. Hams should be resalted every 7 to 10 days during the curing process. Using the second half of your mixture, again rub each piece and cover any bare spots.

Give the salt plenty of time to work. In colder weather, curing time is longer. In warm weather, though, fats may become rancid if the salt mixture is not adequate. Watch your meat carefully and adjust salting times to the individual piece.

After the curing time is completed, wash the meat in lukewarm water. Soak bacon and hams from 30 minutes to an hour. Remove salt and grease by scrubbing, then hang the meat in a cool, dry place to drain. Drying the cured meat may require a week or more.

When each piece is thoroughly dry, wrap it in muslin and several pieces of heavy brown paper, then place it in a heavy brown paper grocery bag to further protect against insects. Hang each piece where it will remain cool and dry.

You may eat cured meat at any time or, if you have a good cool place for storage, it can be kept for a year or more.

CHAPTER 11

ARIES, THE SIGN OF THE RAM

Sheep are ruled by Aries and are also in the sixth house, the House of Small Animals.

Many early societies were based on agriculture, and even today in certain primitive cultures wealth is determined by the ownership of flocks and herds. Flocks have roamed every continent since earliest biblical times. Wherever humans have wandered, from barren Arabia to the pampas of Argentina, from the cold climate of Iceland to the near desert of the Australian interior, breeds of sheep of every description have followed.

Sheep revitalize neglected land, even grazing down and killing off poison ivy and wild blackberries, but they do graze land very closely. In the old days in the Southwest, this was the cause of much trouble between sheepmen and cattlemen.

GETTING STARTED

As a homesteader on small acreage, you would do well to start small, with a few ewes carefully chosen.

You must do some advance preparation, and the first order of business is fencing. You'll need good stock fence about 32 inches high. Fasten this to sturdy posts of wood or metal 2 or 3 inches above ground level, with a strand of barbed wire above and below the fencing. Top this with three strands of barbed wire, each about a foot apart. These last three strands are not intended so much to keep the sheep in as to keep marauding dogs out. Cut timber for posts in the third or fourth quarter.

Sheep equipment, reduced to its simplest terms, is uncomplicated and inexpensive. Even in the North, an open-front shed is usually adequate if lambing takes place late in the season. However, for early lambing, a warm, well-ventilated barn is necessary in cold climates.

With a good lambing barn, temperature is not as important as dryness. Baby lambs can withstand cold weather but not a heavy rain, for a drenching too soon after birth can result in pneumonia and even death.

PASTURE

Sheep are hill animals by nature and will not do well in damp, badly drained pastures. Confinement in such areas is the cause of many a sheep ailment.

Sheep will quickly get rid of weeds, which, incidentally contain many vitamins and minerals. In fire heather and gorse tracts, they encourage the growth of young shoots of tender grasses. They will make good use of such tough foods as reeds, sea grasses, and seaweeds. They can survive in a bleak pasture where other animals would starve, but they will not put on good flesh and wool under barren conditions. They thrive on sedges, rushes, heaths, and mosses, as well as wiry grasses, if found in abundance.

Sheep are greatly benefited by salt air. Organic iodine, in the form of seaweed, helps to supply the crude salt that sheep require to maintain health and remain worm-free. It is also believed that salt given to sheep makes for fine, soft wool, strengthens the animals, and prevents foot rot. Spreading of salt in pastures is recommended in many old farming books, as a preventative against worms and fluke.

If sheep are penned at night, or barned, move them on to fresh pasture each morning. Take precautions against bloat by feeding them hay before you move them. Ordinarily, green pasture causes no trouble, but lush alfalfa, clover, or vetch can bloat them if they are allowed to graze on an empty stomach.

Sheep are very fond of ladino clover and alfalfa, both of which make for good, fast growth.

Ladino clover should be at least 8 to 10 inches tall before you put sheep to graze on it; alfalfa should be a foot or more. Anthrax, the worst sheep killer, can be avoided by not allowing your sheep to graze too closely on sparse, late-summer pastures. Anthrax germs live in the soil and can be picked up by the animals on short grass.

Pasture rotation is the best preventative of worms and parasites. Also make sure that your sheep never have to drink stagnant water. Such water holds the larvae of the land snail, which is host to liver fluke.

TYPES OF SHEEP

Sheep have always been important for their fleece, but breeders have developed sheep primarily for their meat only in the past 200 years.

Sheep are generally classified into four groups, depending on their fleece. These are long wool, medium wool, fine wool, and carpet wool (the carpet wool is of little significance).

Three of the most important long-wool breeds come from England. They are the Lincoln, the Leicester, and the Cotswold. There are many of these in the western United States; they are among the largest of domestic sheep and produce the longest fleece.

The medium-wool sheep are grown primarily for their meat, but are also a source of wool. The most important breeds are the Hampshire Down, Shropshire, Southdown, and Suffolk. These are mostly raised as purebreds but on western ranges they are also used for breeding with the ewes of native sheep to produce lambs for market.

Most fine-wooled sheep originated from the Spanish Merino. They have been greatly improved in the United States, and American Merinos are considered the best in the world. The Rambouillet, another breed descended from the Spanish Merino, is prized for both its wool and its meat.

CHOOSING THE EWE

Look for a straight back and big, wide-set, straight (but not long) "leg-o-lamb" legs. The body should be deep through, from the back down to the stomach. This indicates a good big lamb-carrying capacity. The large blocky ewe is likely to produce more twins, better lambs, and a heavier fleece. Check for a well-developed bag and teats that you (and the baby lamb) can locate easily.

Well-fed ewes at pasture

Try to select a ewe that is 3 or 4 years old. If this is not possible, select one that is a bit older rather than younger. An older ewe is more experienced, will lamb more easily, and should do a better job of rearing her young. (The number of large front teeth tells the story. There are two large front teeth at the end of the first year, and two more are added each year until the ewe has a total of eight when she is 4 years old.) For buying, choose the new moon or first quarter of the moon, on a day marked favorable for your sun sign. The moon should not be aspected with Venus.

Early fall is the recommended time.

BREEDING

A female born between December and March should be bred the following September through January. However, a more mature animal is bigger-bodied and better able to bear and rear healthy lambs, so it is better to wait until the ewe is 15 to 18 months old. Ewes start coming into heat in July and continue every 17 days through January. The gestation period is approximately 150 days, roughly 5 months.

As with other farm animals, young born during the fruitful signs are generally more healthy, mature faster, and make good breeding stock. These signs are the feminine water signs of Cancer, Scorpio, and Pisces. If you are raising lambs for meat, the earth signs of Taurus and Capricorn are good. The lambs will still mature quickly but will produce leaner meat. Lambs for show should be born in Libra.

Sheep like affection, and the lambs are so attractive that it is difficult not to give it to them, but experience has shown that ewes that have been made family pets do not make good mothers. Also, if the lambs are eventually to be slaughtered for meat, you will find it almost impossible to kill a pet.

This does not imply a lack of kindness. Soothing talk, removing briers from their noses or coats, even occasionally rubbing their noses and chins will all help to keep them contented and good producers.

THE RAM

The ram should be in good condition. Feed well a month before you put him in with the ewes. Give him 2 or 3 cups of rolled barley or oats morning and night as a strength booster. If you have six ewes or fewer, it's probably more economical to make arrangements to have them bred by an outside ram, rather than feeding and maintaining one yourself.

It is easier on the ram if two weeks before breeding the ewes are "crutched out" by clipping or shearing soiled locks and excess wool about the dock (tail stump), vulva (under the tail), and around the bag area.

FEEDING THE PREGNANT EWE

The ewe, too, must be in good condition. A month or so before breeding (and toward lambing time) give the ewe small amounts of grain. Well fed, she is more likely to produce strong, healthy twin or even triplet babies. You may feed half bran and half oats, or use corn or barley. This grain feeding should be continued as long as the ewes are nursing, to make the lambs gain weight fast and cheaply.

Put ewes onto good green feed, especially mustard, several weeks before the breeding season begins. Ewes in lamb should always have access to plenty of fresh water, running water if possible. This is one of the most important items in their daily diet. Range that is wide and free is desirable to prevent harmful overweight. This also gives them the advantage of eating a wide variety of herbs, especially the twigs and leaves of wild raspberry bushes.

Raspberry is an important herbal aid to quick and easy parturition. Even animals that have been slow and difficult breeders have had easy and speedy lambings with the addition of raspberry foliage to their diet. If raspberry is not available, briers or blackberry can be used, but neither will be as effective. Successful parturition may also be aided by linseed.

In-lamb ewes are best kept on a light diet. Heavier feeding follows lambing to encourage high milk yield. Kale is an excellent feed for in-lamb ewes. After the ewes have lambed, give them root crops such as carrots and turnips. Oats, rich in fat, also aids a heavy milk yield. Most of the aromatic herbs, lavender, marjoram, rosemary, sage, and thyme, for example, will increase milk yield, as will milkwort and speedwell.

LAMBING

Previous to lambing, clip the ewe around the bag area to make it easier for the baby lamb to find the teats.

You may want to confine your ewe in a 4-by-4-foot lambing pen. This ensures that the lamb stays with its mother. On a fairly large acreage, the ewe may take off to the woods to lamb; in bad or cold weather, this could result in the loss of the lamb. The ewe can get into difficulties if you are not there to help. Occasionally she gets on her back during delivery and must be rolled over or she could die. She may also need temporary aid in standing or walking.

A ewe's milk bag begins to swell 3 to 10 days before she is ready to lamb. The vulva becomes soft, swollen, and wrinkly 3 to 5 days before lambing. The ewe will appear sunken just ahead of the hips about a day or two before she lambs.

As delivery time draws near, she becomes restless and prefers to be alone. Check her every 2 or 3 hours until delivery and for several hours afterward, making sure the placenta has been fully discharged. From the time of breaking water, most ewes will lamb somewhere between 30 minutes and 1½ hours.

If she is in labor more than 3 hours, you may need to help her. Take hold of the exposed toes and pull down. Never pull up; an upward pull will cause paralysis in the ewe. You should also make sure that the prolonged labor is not because the head of the lamb is turned back. If you feel the problem is too complicated for your skill, call a veterinarian.

ONE LAMB OR TWO?

With the older ewe, two lambs or even three are considered desirable, and these are heritable characteristics that sheepmen look for. However, a ewe with her first lamb will not develop as large a bag nor as much milk as she will in succeeding pregnancies. It is wise to allow her only a single lamb to nurse and rear from her first pregnancy.

If she has twins, remove one and bottle-feed it yourself for the first 4 months, or ask your county agent to find a 4-H youngster you can place it with. If you don't have the time for feeding and there is no other possibility, you may have to destroy one of the lambs. However, do not do this for at least 48 hours. If this is a first lambing, one lamb may be weaker than the other and will die. You might just guess wrong, and the ewe is left with no offspring.

After the first pregnancy, you can let the ewe raise all the lambs she brings. Of course one lamb, receiving all the nourishment, will finish out faster, but if you have plenty of pasture, taking a little longer won't really matter.

LAMBS

Ewes well fed and in good condition (not overfat but well fleshed) will bring strong, healthy lambs, active, sturdy, and resistant to cold.

Lamb

Take care that they are not crowded in the pens, for young lambs develop quickly and need ample growing space. They need room for their limbs to grow, nutritious food for their bodies, and good air for their lungs. Lambs love the open air just as their mothers do and when in normal health will thrive even in the hardest frosts.

Lambs need protection only for the first few days; after that, they should not be kept too warm or too sheltered, for they may take chill later on when they are sent out into the open spring fields.

Wide grazing range for the ewes from whom the lambs are feeding is essential to enable them to make the best of milk for their offspring. Wide range is also necessary to ensure freedom from worms.

To keep lambs free of worms, the ewe's milk should be heavily charged with garlic, either by grazing or by feeding garlic in a mash of bran and molasses.

Coughing in lambs may be caused by worms. Difficulties in the respiratory tract are often caused by inflammation. Give garlic with molasses or make a strong infusion of equal parts of sage and thyme to which honey is added. (To make an infusion, pour boiling water over the herbs, allow to steep for several minutes, then strain and cool.)

FEEDING OUT

Using a "creep feeder" to provide extra grain brings lambs to heavier weight more rapidly. Limit this supply, however, to make sure the lambs do not overeat. Consult a veterinarian about vaccinating to prevent enterotoxemia, the overeating disease, if free access to feed is provided.

You may allow free access to milk, grass, hay, and grain for the first 2 or 3 months. Cut the hay or grain from the lambs' diet a week or so after they are on full pasture.

Lambs intended for fall fattening should get some grain again in late summer when pasture grass is usually poor. If you have only a limited amount of pasture available for spring, summer, and fall, feed alfalfa pellets. Sheep relish these and there is the added advantage that pellets keep their wool free of hay. If you decide to use the wool yourself, this will eliminate a lot of cleaning.

Shear wool to increase growth in Cancer, Libra, or Taurus and in the first or second quarter. It is also satisfactory to shear in Leo, Virgo, Sagittarius, and Aquarius.

CASTRATION AND DOCKING

These two important steps should be taken 7 to 14 days after a lamb is dropped. Always have a good disinfectant on hand to use as an antiseptic on equipment.

Lambs are born with a long tail, which collects droppings and makes them susceptible to maggot infestation. An elastrator (available from most farm stores) with rubber bands is the simplest method of docking. The amount of tail cropped depends on your preference. Some ranchers like to crop as close to the body as possible, others leave 1 to 3 inches of tail.

Ram lambs, unless kept for breeding, should be castrated before docking. For these operations, select a date when the moon is not in Virgo, Libra, Scorpio, or Sagittarus, and when it is within 1 week before or after the new moon.

SLAUGHTERING

Ram lambs should be slaughtered when they are between 5 and 10 months of age. A good choice for well-grown animals is 6 months. You can do this yourself or have it done. Hang the carcass 1 to 3 days in a cool place, after which it should be cut, packaged, and frozen. We have done this ourselves, laying the carcass out on a large table, thus saving considerable expense. Usually a lamb at this age will dress out to between 35 to 75 pounds, depending on breed and feeding conditions. Organically raised lamb is a tasty treat you cannot possibly imagine until you try it yourself.

In addition to the generous amount of high-quality, readily digestible protein supplied by lamb, it is also a rich source of iron (for building and maintaining red blood and prevention of anemia), as well as the crucial B vitamins, including thiamin for healthy nervous system, riboflavin for good vision and clear eyes, and niacin for healthy skin.

MANURE

A fully mature ewe produces nearly a ton of manure a year. The sheep, like the cow, is a ruminant, having four stomachs instead of one. It is able to thrive on bulky forage and roughage, even ground corncobs and their stalks.

Supplements (soybean meal, molasses, bonemeal, salt, and cod-liver oil) sprinkled on roughage is desirable, not to feed the sheep, but to feed the bacteria that thrive in the rumen. These bacteria break down the coarse forage into high-nutrient elements.

All of this adds up to a particularly rich "hot" manure. Use it sparingly and always compost it with other organic matter before you incorporate it into a gardening area. Sheep manure is a good source of potash. Use the decreasing moon to spread organic fertilizer.

SHEEP PROBLEMS

Sheep, in common with most domestic animals, may be attacked by many species of worms. It is quite useless to dose animals for worms and then let them remain in a pasture that is infested. Before the land is used again as pasture, cleanse it by liming, by growing a heavy crop of mustard and plowing it in, or by planting it with garlic. As with other farm animals, worm treatments are best given when the moon is waxing and near full.

Garlic, aided by molasses, is one of the best worm remedies. It has great powers of pulmonary penetration, important because the lungs of sheep (as with goats) are a common area for worm infestation and very difficult to reach and cleanse.

LAMBSKINS

Even if you don't want to do your own spinning, you will take great joy in using properly tanned lambskins. These are a real luxury item, beautiful, practical, and readily saleable.

Garlic is also an important preventative; sheep allowed to pasture on the wild vari-ety, which grows abundantly in many parts of the world, are seldom subject to worms. The sheep love it and will graze it avidly when available.

Apply a strong solution of garlic leaves or roots into the sores caused by para-sites. Garlic taken internally disinfects the bloodstream and repels skin parasites. Dusting with quassia chips is also effective against external parasites.

The best tonic for sheep is considered to be molasses and the best general med-icine linseed oil, which is a gentle purge as well as a good conditioner. Carrots, too, make an excellent tonic, and are of value in preventing or curing infections.

Foot rot is usually caused by pasturing on damp, low land and by over-rich feeding. The natural home of sheep is on stony, dry hillsides with sparse vege-tation. Exercise over hard ground keeps the feet firm and in good condition.

Treat lice and ticks by dipping in a strong solution of derris, preferably pow-dered. You can also add a little eucalyptus oil or camphorated oil to the dip.

LOCOWEED

Locoweed is any one of several kinds of perennial herbs that grow in western North America. It gets its name from the Spanish word *loco,* meaning "crazy," because of the strange actions of animals poisoned by it.

There are about 100 kinds of locoweed, but many are not known to be poi-sonous. Three of the more common varieties are the white, purple, and the blue locoweed. (They are named for the color of the pealike flowers.) The plants have erect or spreading stems, each with many leaflets.

The symptoms of poisoning vary somewhat among horses, cattle, and sheep. Horses become dull, drag their legs, eat infrequently, and lose muscle control. Soon they lose weight and eventually they die. Cattle react in much the same way, but sometimes they run wildly about, bumping into objects in their path. Sheep react more mildly to the poison. Animals raised on a range usually do not eat locoweed when other food is available.

Ranchers destroy locoweed by cutting the roots about 2 inches below the surface or by spraying the plants with herbicide (ask your county agent for rec-ommendations). Destroy plants when both the sun and the moon are in barren signs, and the moon is decreasing, fourth quarter preferred. Barren signs are Leo, Aries, Virgo, Aquarius, Gemini, and Sagittarius, listed according to the degree of nonproductive qualities.

CHAPTER 12

NEPTUNE RULES PISCES

And fishes are ruled by Pisces, which is known

in astrology as the Sign of the Fishes.

People were enjoying the sport of fishing long before Izaak Walton wrote his famous book and still are. Fishing is important both for food and for fun: Some people depend on it for their livelihood and others for their peace of mind. The desire to "get away from it all and go fishing" is still with us.

A FISH'S WORLD

Fish see with eyes much like ours, modified for vision under water, of course, but most cannot see very far. They depend instead on their sense of smell. Their smelling organs are large, leaflike structures located in a pit that has one or two nostrils.

Fish do not have a keen sense of hearing, but sound is not important in their lives. They can hear only low-pitched sounds.

Fish have a sense of touch throughout the skin. They may also have a sense of taste on the outer parts of their bodies. Catfish are said to taste with the outside of their bodies although their main taste organs are in "whiskers" near their mouths.

Strangely, fish do have something that may be regarded as a special or sixth sense. It is located in the *lateral line,* a long row of tubes and pores over a nerve running along the side of the body from head to tail fin. This sense organ feels vibrations too low to be heard by the human ear and can detect the footsteps of a person on the bank of a stream. Experienced anglers know that a footstep is more likely to frighten fish than is the human voice.

FISH AND COLORS

Some fish may be able to see and distingush certain colors. Perch, trout, minnows, and others have been proved by various tests to recognize a fair range of shades. Perch regularly fed on larvae previously red-stained later were easily deceived by red wool. Similar tests have been made with food dyed green, yellow, orange, and brown, with the same results. Quite probably those species of fish that change color to match their surroundings can see these and perhaps other colors. All of this makes it easier to see why expert anglers are so careful when selecting baits.

WHEN TO FISH

A great deal of fishing lore has passed down from those who understood it well, and much of this lore is based primarily on timing and weather conditions.

During the summer months, fishing is considered to be best from sunrise to two or three hours after and later in the day from about two hours before sunset until one hour after.

During the cooler part of the year, fish do not bite actively until the air is warmed by the sun. The best hours seem to be from noon to 3 o'clock.

Warm and cloudy days are good for fishing. Fish come closer to the surface at this time because the water is warm and the sun does not hurt their eyes. (Fish have no eyelids.) The best water temperature is between 55°F and 74°F. The clearer the water, the better, and preferably with a slight ripple.

The most favorable winds are those from the south and southwest, and any offshore breeze. Easterly winds are considered unfavorable.

The best days of the month are when the moon changes quarters and particularly if the change occurs on a day when the moon is in one of the watery signs, Cancer, Scorpio, or Pisces. The best day in any month is the day after the full moon. A high barometer always helps, too.

If you are fishing in the ocean, the best time is one hour before and after high tide, and one hour before and after low tide. Low tides are considered to be halfway between high tides.

Solunar tables, giving the best times for fishing, are widely available on the Internet. These determine when fish (and game) are going to be most active on any day of the year, and whether they are going to be active at all on that particular day.

These tables are not "magic," nor "instant fishing," but rather a valuable tool. There are many reasons why a fisherman does not catch fish, including the wrong bait. Sudden cold fronts, too heavy a line, and just plain fishing where fish are not present can bring about an empty reel. Nothing can cure all of these things, and the solunar tables do not attempt to. What they do is to accurately predict the tides of each day when you are most likely to catch fish.

Observe your local weather conditions in planning your day. I know, for instance, that a heavy early-morning fog, especially in spring, usually forecasts a fine, bright, sunshiny day. On the other hand, especially in the polar airstream behind a depression, clear skies in the early morning may be followed by cumulus and, later, cumulonimbus clouds with showers, which leads us to the saying: "Shiny morning, cloudy day; cloudy morning, shiny day."

FISHING TIPS FROM LUCY

My friend Lucy Hagen, a champion fisher lady, once told me something that her father, Ole Hagen, had taught her: Watch and see if the cattle are grazing when you plan to go fishing, because if they are, the fish will be biting too. The reason for this is, of course, that the wind tides governing the fish also govern land animals. To the student of astrology who observes all nature, the message is plain.

Another bit of fishing lore Lucy shared with me was this: Rub a worm (real or rubber) *lightly* with oil of anise to attract bass.

Lucy fishes with these worms on the lake bottom because experience has taught her that the largest bass are down there in the deep water. She rigs the line with a slip sinker and very gently works it across. She became so experienced that she could just pull this line across the area where she was fishing and tell much about it from the "feel."

Lucy once showed me the bait she used primarily for bass fishing: rubber worms in every conceivable color, even striped and polka-dotted. Her first choice is a deep purple worm; for some reason, most of the fish she has caught have been on this color. She uses this for the greater part of the year but during the winter months she likes a black one.

Lucy also likes to use the "toothpick method." As the rubber worm is threaded on the hook, she thrusts a toothpick through the eye of the hook and then clips it flush with the worm on either side. This, she told me, prevents the loss of the worm and keeps it from being pulled off the hook, mangled, or destroyed.

Like other good anglers, Lucy considers certain areas better than others. "Look for an area with high and low places," she told me, "for instance, where a small stream may be entering a lake and has cut a deeper channel." She also watches for schools of smaller fish: "Big ones will be there feeding," she says. Lucy considers brushy places good, either natural or where pines or cedars have been placed in the water.

She then talked about the effect of temperature on fishing. "In the hot summertime," she said, "right before a cool spell moves in, the bass will go on a real wild feeding spree and the same thing holds true in the winter when a warm front moves in."

I asked Lucy about the effect of moonlight on fish. "When the moon is full," she replied, "and casts its light upon the water, fish down the moonbeam for the best results."

CALLING UP EARTHWORMS

Bluegills and perch are usually hungry the last few weeks before freeze-up in November, and earthworms are their favorite bedtime snack. So on a bright fall day, pick a moist, humus-laden spot and drive a stake about 5 inches in the ground. With the back of the hammer, rub the wood so that it gives off dry grunts. The curious or frightened worms will obligingly come out of the ground.

KEEPING FISH BAIT ALIVE

Keeping live fishing bait cool is a major problem for many anglers, all the more so in hot weather. In a matter of minutes, critters like minnows, crawfish, grasshoppers, worms, and crickets can become overheated and die.

Minnows are the most vulnerable, especially if transported some distance. Some ice cubes and a plastic bag works wonders for them. Punch several small holes in the bag, fill it with ice, and secure the top. Place the bag over the minnow bucket. Cool water will drip into the bucket as the ice melts and thus keep the water from overheating. If ice isn't available, wrap a piece of burlap around the bucket and wet it. Evaporation will do the cooling job. Arriving at your destination, don't just plop the bucket into the water. Surface water, if warm, causes the minnows to go into shock if suddenly transferred from cool water, and they will die immediately. When transferring, try "tempering," that is, take several minutes to mix warmer water with the cooler water in the bucket, giving the minnows time to become acclimated to the temperature change.

Keep crawfish alive by placing a tray or two of ice cubes in the bottom of the bucket, then covering them with several folds of burlap topped with a thick layer of grass. Dump the bait onto the grass and keep the bucket out of the sun.

It's a simple matter to keep worms alive, but it's also easy to mistreat or accidentally destroy them. Put them in a can with some soil and mulch such as grass clippings, and place the can overnight in your home refrigerator. When you go fishing, quite likely you'll take along an ice chest for cold drinks. Simply tie a plastic bag around the worm can, leaving a short loop of string. Hook this outside the cooler and let the can hang inside, suspended over the ice and cold drinks.

Keep crickets and grasshoppers alive the same way. The cold slows down their movements and makes them easier to put on the hook, too.

A little ingenuity and a handful of ice will solve most of your bait problems during warm weather, and that will help you keep your "cool" when you go fishing.

FISH LORE

Black bass (largemouth). Native Americans call this one the bulldog. Chubby, stout, and well muscled, it will fight with dash and spirit. It likes to roost under lily pads. Indians cast for them using long poles among weeds and lily pads. On the hook is attached medium-size frogs. Smaller bass are caught with minnows and worms.

Black bass

Carp. Fish near the bottom using a small hook and doughballs.

Catfish. Use angleworms and chunks of liver as bait. The more stinky baits you use for catfish, the better. They like small frogs and parts of larger frogs, or try doughballs, chicken entrails, crawfish, grasshoppers, and minnows.

Muskellunge. The muskie is known as the king of freshwater fish, it is a savage striker. Trolling is a good way to catch them. The bucktail, with a large hook concealed in a portion of the tail, is the bait. Other baits may be live suckers and chub minnows, small perch, field mice, strips of fish, and frogs. Play your fish out before trying to land him; he is a wily fooler.

Yellow perch (white perch). You can catch them at almost any hour of the day. Bait with minnows, worms, grasshoppers, grubworms, crawfish, strips of perch belly, and even their eyes! They're good to eat but hard to scale.

Great northern pike. This pike feeds early in the morning and late in the afternoon, which is when you'll find it in weedy and shallow water. At midday look for them in deeper waters. Entice a great northern as the Native Americans do with shiner, chub, or sucker minnows measuring about 4 inches long. Always hook the minnows lightly in front of the dorsal fin.

Great northern pike

Sheepshead. This fish is good eating but a bit rich. Native Americans use angleworms for bait. Fish near the bottom using heavy sinkers, no bobber, and a smaller hook. These fish are fighters.

Smelts. For these sweet small fish, bait your hook with small minnows, crawfish, or parts of other fish.

Walleyes. Walleyes seek shallow water in early spring, over sandbars, rocky reefs, and swift waters. You can take them on shiners about 2½ inches long with a spinner attached and dropped to within a foot of the bottom. Native Americans refer to this as still-fishing. Jerk the hook slightly at intervals to draw the fish's attention. They bite on minnows, but they take their time, so be patient before setting the hook.

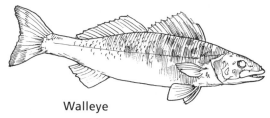

Walleye

TO FREEZE FISH

For fish to taste really good it must be fresh. If at all possible, freeze fish on the same day that it is caught. Handle it carefully and keep it in the refrigerator or in cracked ice to prevent rapid spoilage. This is particularly important in warm weather.

Fresh fish has firm, elastic flesh that resists being indented when preserved. The eyes are clear, moist, and full; gills are bright red; skin is shiny. The scales of fresh fish adhere closely to the skin.

Fish should be prepared, ready to cook when you remove it from the freezer.

Scrape off the scales, using the back of a heavy knife. Make a slit down the body cavity from gills to vent; remove entrails and scrape the backbone clean. Remove fins, gills, head, and tail; then wash the fish thoroughly inside and out. Fish may be left whole or cut into steaks or boneless fillets.

Immerse steaks and fillets of lean fish in a cold salt solution called a brine dip. To make this brine, dissolve ½ cup salt in 1 quart water and immerse fish for 20 seconds. This will reduce leakage or drip when the fish is thawed. Do not brine fatty fish.

Fat fish such as lake trout, pink salmon, and mackerel can be kept in frozen storage longer if they are dipped for 1 minute in a solution of 1½ teaspoons ascorbic acid to 1 quart cold water.

Fish tends to dry out quickly in storage, so protect it with the best vaporproof containers and wrappings. Place double thicknesses of freezer paper between each fish, fillet, or steak so they can be separated before they are completely thawed.

I often save containers, such as those in which cottage cheese is packaged, place the fish therein, and fill with water. It takes longer to thaw, but the fish keep longer for me this way. Fish should always be frozen in meal-sized amounts and kept at 0°F.

Lean fish, or fat fish treated with ascorbic acid solution, should not be kept in storage for longer than 6 to 8 months.

FRESHENING FISH

Flavor exists only in the skin of fish and becomes communicated to the body only on being boiled. It is sufficient, before dressing fish that are taken from ponds or stagnant waters, to put them alive into a tub and pour fresh water over them. Add half a handful of salt, stir it up well among the fish, and finally rinse them with fresh water.

I have found it possible to freshen the flavor of fish kept overlong in the freezer by soaking it for an hour or two in saltwater before preparing it for the table.

FISH AND YOU

Fish is a great natural food, good for body-building and developing muscular strength. It is the easiest of all flesh food to digest, if it is not fried. Ocean fish are a good source of iodine and particularly helpful to those in a rundown condition or suffering from glandular disorders. Fish contain the vitamins B_6, B_2, B_{11}, and D as well as nicotinic acid.

My parents insisted that I eat my fish, telling me that it was "brain food." I thought this was very funny, not realizing until much later in life that they were correct. Fish, along with whole wheat, egg yolk, milk, nuts, beans, and peas, is especially rich in phosphorus. In the human body, phosphorus compounds are found chiefly in the bones, the brain, and the nerves.

FISH FARMING

If you have sufficient land and decide fish farming is for you, this may well become an enterprise that affords you both profit and pleasure. A fishpond is a delightful part of a farm or ranch, adding beauty to the land and also providing recreation for farmers, ranchers, their friends, and possibly even paying guests. On a suitable site a pond makes good use of the land, and the impounded water can be utilized in several ways. In the South and Southwest, in particular, where natural rainfall is often inadequate, a good pond provides water for animals to drink and irrigation for the family garden.

Managed well, fishponds can be quite profitable. To produce the most income and recreation, of course, they must afford good fishing. Disappointments usually result from mistakes in construction, stocking, or management. In my *Catfish Ponds and Lily Pads,* you may find much to abet you in your quest for a fine fishpond. (See Suggested Reading.)

FISH AS PETS

Tropical fish are extremely beautiful, graceful to watch, and soothing to the nerves. As many doctors with aquaria in their offices will attest, the presence of a well-established fish environment does more to quiet an apprehensive patient than a soft chair or a reassuring word. My own dentist uses a former gasoline pump as an aquarium, and it is both unusual and decorative.

Goldfish are probably still the most popular beginner's fish, having been domesticated since the 10th century in China. But many small freshwater and even saltwater varieties are suitable for home aquaria; they require relatively little space, are not costly to maintain, and on the whole are odorless, noiseless, and entertaining.

Tropical fish are for the most part diurnal (awake during the day), and their behavior is easily viewed in a glass tank, where they may thrive and even breed, affording the observer a chance to study the entire life cycle. In addition to goldfish, popular freshwater species include guppies (from the Caribbean), medakas (from Japan), fighting fish (from Thailand), and swordtails (from Mexico and Central America).

The best container for fish is a glass tank, with a stainless-steel frame made watertight by a thin layer of aquarium cement, which allows for changes in water pressure and temperature. A cover is necessary to keep the fish in and foreign matter out; screening is probably the best material for this. The water in the tank must be specially conditioned or aged in order to remove toxic elements before fish can live in it. Usually tapwater can be conditioned by letting it stand exposed in shallow nonmetal containers, by having compressed air forced through it, or by introducing aquatic plants.

Overfeeding is the cause of more aquarium failure than any other error. Unused food will rot and foul the water, as well as stimulate the plants to a dangerous degree. If you think you are prone to feed excessively, consider purchasing some scavengers, which also help to keep down growth of algae. Chief among those most often used are snails, the pond snail *(Physa),* for example. They multiply rapidly, but you can crush young pond snails between the fingers, thus providing a fine diet item to all fishes consuming them. Others are the red ramshorn or coral snail *(Planorbis corneus* var. *rubra),* a highly developed color phase of the brown ramshorn snail, which is also decorative.

A fish collection will grow larger if you (or the fish) have success in breeding. Fish reproduce in several ways. Most are egg layers and scatter their eggs, either in clumps (goldfish) or individually (zebrafish); carry their eggs (medaka); or build nests (fighting fish and paradise fish). Some bear their young live (swordtails, playing fish, guppies, and mollies).

Fish occasionally do get diseases, so you might want to bone up on symptoms and home remedies. However, you may need expert assistance. If a fish seems ill or injured and you can't get help, you are probably better off removing it from the tank and disposing of it before it can hurt the other inhabitants; dead fish should be removed immediately.

Keep a close eye on your fish and you will begin to realize a great many interesting things. Not only are they a great source of pleasure and interest but some fish also learn their feeding routines and may even recognize the human being (or whatever they think it is) at the edge of the tank, especially if you tap the glass. Remember that fish, even the tiny decorative ones, are still fish. Fish depend heavily on the sense of smell and their hearing is not keen, but they are extremely sensitive to vibrations.

It is even possible to train fish to move to certain locations or in certain patterns if you are patient and use food as a reward.

Keeping fish as pets can be a most enjoyable activity, adding beauty, interest, and movement to any room.

MEDICINAL HERBS FOR ANIMALS

Herbs used to maintain health or treat disease should

be harmless, gentle, and nonviolent in their effects.

Today we are experiencing renewed interest in medicines derived from natural products, along with a revival of interest in natural foods. In fact, the bounty to be obtained from the vegetable and animal kingdoms has hardly been touched, though humans have made medicines from them since ancient times.

ANIMALS AND HERBS

Once animals roamed free and could pick and choose as nature intended. In fact, it is a common belief that early humans learned the art of natural medicine by observing the natural instincts of animals.

Our domestic animals can no longer range freely, so we must plan for them, as well as for ourselves, in making herb tonics available. The best way to do this is to plant the helpful botanicals in pasturelands and alongside hedgerows.

Place deep-rooting herbs in grass mixtures and sow the seeds together. As they get established, they will reseed themselves.

If you grow your own herbs in garden, pasture, or hedgerow, you can be sure they are not sprayed. If you gather them elsewhere, make certain they have not been. They will do more harm than good if loaded with chemicals.

GATHERING HERBS

✿ Gather for preservation during the time when each species is at its best stage of growth, usually in spring and summer. Gather in the morning after the dew is dried but before the hot sun of noon causes wilting. Do not gather wet herbs; they may turn moldy.

✿ Roots to be preserved should be taken before the sap rises in the spring, or in the autumn after the leaves are shed.

✿ Gather leaves when they are small and just unfolding. Yellowed or faded leaves lack medicinal powers and should never be used. Snip off stems and use them in your compost heap.

✿ Gather flowers when they are just opening, if possible before they are visited by bees or other insects. Flower buds, because of their high moisture content, may be difficult to dry, but they have value for use fresh. Never use faded flowers. Snip off stems; keeping them prolongs drying time.

✿ Seeds may be left on the plant almost to the point of shattering. Often the plant itself indicates that the time for gathering is approaching by the yellowing of its leaves.

✿ Barks, like roots, should be gathered either in spring or autumn. At this time damage to the trees is less likely as well. To avoid damage, try to take bark from several different trees, cutting just a small amount from each.

PRESERVING HERBS

Spread herbs on shelves or tables and turn them frequently. To make a good drying table, prop an old window screen between two chairs. This permits air circulation both above and below.

Do not use heat from a fire. Quick drying destroys medicinal powers. Herbs should dry slowly and naturally. Avoid direct sunlight. Be patient; it may take several weeks for the moisture to evaporate from your leaves and flowers.

Seeds and fruits usually require a longer drying time. When the weather cooperates, sun-dry them on plant or tree. These can then be gathered and stored directly. Dry off larger fruits slowly in a gentle oven, if necessary. Sometimes they may be sliced and smoke-cured.

Roots and barks require slow and careful drying. They must also be cleansed of dirt and other accumulations. Store them in paper sacks or even in cotton bags. An old pillowcase, much washed and porous, makes a good container.

Never suffocate your herbs by putting them in transparent plastic bags! Put them in old-fashioned brown paper bags and hang them out of the sun in a warm, dry, airy place to preserve their best healing qualities and strength.

Keep a close watch on all your botanicals lest they spoil or be devoured by rodents, moths, beetles, and other insects.

Your herbs, when carefully dried and preserved, should maintain their powers for about two years. Don't throw unused herbs in the trash bin; put them in your compost heap, where they will still be of value as they decompose.

DOSAGE

In treating animals with herbal medicine, exact dosage is not always necessary. The quantity of medicine varies with the breed of animal, of course, the type of illness, and its severity.

Generally spreaking, a heaping handful of herb, whether leaf, flower, seed, bark, or root, should be brewed in a pint of water. Many animals will lap up this brew with gusto, enjoying it tremendously if it is sweetened with honey or molasses. This herb tea is useful during the winter months. During the spring and summer, I find it preferable to feed most of the herbs fresh gathered and raw.

POULTICE

To make a poultice, place your herbs loosely in a flannel bag large enough to cover the area to be treated. Pour boiling water over the bag and then wring out the excess moisture inside an old towel. Use the poultice as hot as possible (but not so hot that it will burn). Bind in place with lengths of old sheeting. This is particularly good for painful joints and muscles.

SALVE

Choose herbs according to the particular ailment being treated and cut them very fine. Prepare a small bowlful of herbal brew, bring to a boil, and add enough powdered slippery elm bark to form a thin paste. Spread this paste, while still warm, on a cotton or linen bandage (again, use old sheeting or a pillowcase), and apply over the wound or affected area.

HONEY

A soothing drink for all animals that reduces internal inflammations is made of honey and barley. Add ¼ pound barley to ½ gallon water. Boil slowly until quantity is reduced by half. While still warm, add ¼ pound honey. If fever is suspected, add 4 teaspoons of finely chopped sage to the boiling water and barley.

BOTANICALS

Medicines made from herbs were the remedies employed by medical astrologers from time immemorial. The planets in their order from the moon to Saturn are each known to have affinity, or sympathy, with certain herbs. Therefore, various herbs are said to be ruled by certain planets. Saturn, for instance, is said to rule such herbs, roots, and barks as amaranthus, aconite, barley, barrenwort, and comfrey. The almond is under the dominion of Jupiter, the lotus in the domain of the moon, the walnut under the dominion of the sun, and so on.

Alfalfa, sometimes called lucerne. Widely cultivated as a fodder crop, alfalfa contains large amounts of protein, minerals, and vitamins. It has small, gray-green foliage and whitish to pale purple, pealike flowers blossoming in early spring and late summer. Alfalfa is nervine and tonic, and you may permit your animals to eat the cured hay at will. It is excellent as a kidney cleanser.

Plant alfalfa in the first or second quarter under Cancer, Scorpio, Pisces, Libra, Taurus, or Sagittarius.

Alfalfa has 10 times the mineral value of various grains. The roots may go as deep as 125 feet and bring up vital minerals unattainable by other vegetation. It is a rich source of the antihemorrhagic vitamin K, as well as vitamins A, C, and E. It is a blood builder, good for teeth and bones, and excellent for milk-producing animals.

Aloe (herb of Jupiter). This plant grows in dry, sandy soils in warm climates. The leaves are spiny, the white flowers inconspicuous. The juice may be expressed from the leaves and sun-dried for another time. The plant is used both internally and externally. Aloe juice used internally treats constipation, indigestion, and worms as well as ulcers and urinary ailments. Externally it is a healing agent for burns, ulcers, cuts, scratches, and innocuous insect bites. Aloe leaves in drinking water are beneficial to sick chickens.

Anise (herb of Mercury). Anise grows well in moist places and is often found in hedgerows. A tall plant, it has feathery leaves and puts forth umbels of creamy flowers. The plant is strongly aromatic, and both seeds and oil are used. Dogs like the scent of anise, and it is said to calm horses when rubbed on their noses. Rub some on bait to attract fish.

Anise contains sodium. It is useful as a remedy for digestive ailments, including colic in young animals. Give about 1 handful of seeds daily.

Anise hyssop, sometimes called fragrant giant hyssop (herb of Jupiter). This plant blooms from early summer until frost. Bees consistently work the flowers from daylight until dark. It has been estimated that an acre of anise hyssop might be able to support 100 hives of bees.

Asparagus (herb of Jupiter). Asparagus is familiar as a cultivated garden vegetable and grows wild in many parts of the world. Horses and cattle are fond of it. The entire plant powerfully influences the urinary system.

Asparagus is useful against obstructions of the bladder and kidneys. The young shoots are best; they are both diuretic (promoting urine) and aperient (gently laxative). Depending on the age of the animal, give 1 or 2 handfuls twice daily. Eaten by nursing animals, however, it may lessen the milk yield.

Barley (herb of Saturn). This cereal grain should be planted in Cancer, Scorpio, Pisces, Libra, or Capricorn; the first or second quarter is the best time. Stretches of wild barley are indicative of lime-rich earth.

To make barley water: Using equal amounts of water and barley, pour boiling water over crushed barley grains. Let stand overnight, then add the same amount of tepid water. Add 1½ teaspoons of lemon juice and 1½ teaspoons of honey to each cupful of barley water, obtained by straining the mixture. Barley water is useful as a drench (drink) for feverish animals.

Beech (herb of Saturn). One of the most widely distributed trees in our country, beech is also one of the most useful and beautiful in any forest. The whole tree is highly medicinal, including the buds, leaves, bark, and nuts.

A brew made from the dried leaves or bark is useful against ailments of the liver and kidneys. Externally, a brew of the buds or leaves is softening to hard wounds.

Birch (herb of Venus). Birch is particularly good to grow around the edges of compost and manure piles, as a secretion from the roots seems to encourage fermentation. Birch leaves are beneficial to exhausted soils.

Extract the sap and collect it by boring holes in the tree in early spring before the leaves appear. A little oil poured on the surface will keep it fresh for many months. Mix 4 tablespoons of the sap with bran and give as a tonic. (Small twigs and inner bark may be used instead of the sap.)

Birch is useful in treating digestive ailments and general debility. The leaves are cleansing to an animal's system and will expel worms. Give at the time of the full moon, when worms are most active.

Boneset (herb of Saturn). Boneset is found throughout the United States in low ground, marshes, and swampy places. Growing 3 to 4 feet tall, it has hairy, opposite leaves and bears large heads of white flowers.

Boneset is a tonic stimulant, promotes digestion, and strengthens the viscera. Made into a tea (using the dried leaves) and taken warm, it induces perspiration as a treatment for colds.

Borage (herb of Jupiter). A plant of fields and woodland, borage has rough leaves and whorls of brilliant blue, wheel-form flowers.

Borage has powerful tonic properties and is an excellent pasture herb. The whole plant is edible. It will increase the milk flow in herd animals. Useful externally, borage can be made into an eye lotion and is also used as a remedy for ringworm. Make a poultice with the leaves to soothe inflammatory swellings.

Honeybees are attracted to borage and prefer it to all surrounding flowers.

Bramble or blackberry (herb of Venus). Bramble grows in hedgerows and woodland. It has thorny foliage and stems, and white, single, roselike flowers. The fruits are avidly eaten by all animals.

To prepare an astringent potion, boil 1 ounce of the root in 1 to 1½ pints of water or milk. Give 1 teacupful every hour or two to alleviate diarrhea (½ cup for small animals). If fresh root is used, a larger amount is required.

Blackberry is an herb tonic in pregnancy. Externally it can be used for skin disorders, helpful to all types of eczema. Warm fresh-picked leaves before applying. The green upper side of the leaves is soothing to the skin; the white underside will draw when applied. Pulped leaves are good for burns and blisters.

Burdock (herb of Venus). Though not "official," burdock is considered diaphoretic (producing perspiration), diuretic (promoting or increasing urine), and alterative (gradually restoring healthy body conditions).

Burdock is a biennial plant with large, rhubarblike leaves. It often grows along roadsides. Animals will not graze this herb, but the roots, sliced and bruised, are a fine blood cleanser. Crushed leaves, externally applied, are a good remedy for scabies and ringworm. A treatment for burns is made from the fruits.

A poultice of the leaves provides relief from bruises, tumors, and other swellings. Some authorities even mention this as an antidote for snakebite.

Caraway (herb of Mercury). Caraway grows best in full sun in medium-rich soil. Plant caraway for an increased milk supply in dairy cows and goats. Caraway seed adds an aromatic flavor to hard breads and also makes them more digestible.

Carrot (herb of Mercury). Carrots are common in the vegetable patch. The carotin principle found in them is useful in eye disorders. Carrots are good for all animals, especially horses, rabbits, and cows, but have particular value in expelling worms in goats.

Chamomile (herb of the sun). This is a low-growing, pleasantly strong-scented herb of waste places and hard, stony ground. It has daisylike flowers with white petals and a yellow center.

Chamomile

As an internal medicine, the flower heads are dried and used in infusions, decoctions, and as extracts of pure oils. Externally, they make a good poultice. Chamomile is useful for aches and pains as well as in the treatment of skin and blood disorders.

Scour in calves may be relieved by chamomile tea. Pick blossoms of the plant on a bright, sunny day and gently place them in a shallow glass bowl of water — pure spring water, if possible. (Blossoms may also be carefully dried and kept for future use.)

When the surface of the water is completely covered with the floating blossoms, leave the bowl in the sun for three or four hours. Then carefully whisk them out with a twig and pour the water into bottles. Add a small amount of brandy, if you wish, to act as a preservative.

A small amount, taken with milk or water, is all that's necessary.

Chickweed (herb of the moon). This creeping plant of waste places has tiny, white, starry flowers. It is rich in copper and of value to all grazing animals.

Lambs, however, should be prevented from overgrazing, which may result in a digestive upset. Chickweed has many of the same soothing and tonic properties as slippery elm.

Externally, chickweed is useful as an eye lotion. You can also make it into a salve for rheumatic inflammations and stiff joints. Incorporate 1 cup of finely cut herb in 2 cups of simmering vegetable oil and heat until the herb is well absorbed. Strain and pour into jars.

Cleavers (herb of the moon). A plant of slender form and climbing nature, cleavers has hairy stems and foliage. Its hooked bristles cling to everything. Cleavers is rich in minerals and silica, which give strength to the shells of eggs and the hair of animals. Although all animals eat cleavers, poultry actively seek it; hence its other name, goose grass. It is also called knotgrass, bindweed, and beggarweed.

Besides its tonic properties (a decoction of the plant mixed with oak bark is a substitute for quinine), cleavers is effective against skin diseases such as eczema, and against abcesses and tumors.

Clover (herb of Jupiter). Plant in the first or second quarter under the sign of Cancer, Scorpio, or Pisces. An excellent pasture and bee plant, red clover has a demulcent action and thus is soothing to coughs. It is good for skin eruptions when applied as a poultice made from the bruised flowers, steeped in water for 2 or 3 hours. A poultice of the flowers is also recomended for ulcers. White clover is also used externally to heal sores.

Comfrey (herb of Saturn). Also called healing herb, knitback, and ass-ear, comfrey grows about 2 feet tall. Its leaves are large and broad at the base and lancelike at the terminal. The fine hairs on the leaves cause itching. Flowers are white to purple.

White clover

Make a poultice by beating the plant to a pulp. It is useful when applied to sprains, bruises, and swellings. The healing substance in the plant has been identified as allantoin.

Comfrey has the power to aid the body in quickly uniting fractured surfaces, hence its other name, knit-bone. For this purpose, use the leaves and roots.

Comfrey is much enjoyed by sheep, cows, horses, and rabbits, which particularly relish the young shoots.

Dandelion (herb of Jupiter). The dandelion herb is blood-cleansing and tonic and is supremely effective in curing jaundice. The leaves strengthen tooth enamel and the white juice helps to dissolve warts.

Goats will graze dandelions and horses enjoy the leaves when they are cut and mixed with bran. Dandelion leaves are a good conditioner for all animals.

Dill (herb of Mercury). This plant, which has fine feathery leaves and pale yellow flowers, likes to grow in hedgerows and on wasteland. The seeds contain an essential oil and the whole plant is very aromatic.

Dill increases milk yield, especially in goats, and is useful as a treatment for all digestive ailments. The seeds are of greatest value but the leaves may also be used.

Dock (herb of Jupiter). Dock is a spring green that many consider superior even to spinach. It has curly leaves and a warm brown seed stalk. The whole herb is antiseptic, and both leaves and root are use externally for fevers, skin inflammations, and to cleanse and heal wounds.

Elm, slippery (herb of Saturn). A very valuable medicinal tree, slippery elm is also known as the red elm and the moose elm. The tree can attain a height of 75 feet. It has reddish bark and dark green leaves, hairy beneath and rough above.

Slippery elm was widely used by Native Americans against diarrhea and other conditions caused by an irritated stomach or intestines. It is also effective for coughs due to colds and for skin irritations.

Stir 2 teaspoons powdered bark into 1 cup milk or water. Add 1 teaspoon honey. (Increase the dose according to the size of the animal.) This is a good treatment for scour.

Fennel (herb of Mercury). Grow fennel in full sun. It likes medium-rich, well-drained soil. Its leaves and seeds are anise flavored. Fennel not only increases milk yield but also imparts a sweet odor to the milk.

Fennel possesses antiseptic and tonic properties. The foliage may be brewed and used for an eye medicine.

Garlic (herb of Mars). Plant in the first or second quarter under Scorpio or Sagittarius. Garlic is widely cultivated but also grows wild. Few plants are so well known and universally used for their medicinal properties. Highly antiseptic, garlic juice is sometimes applied to wounds. Garlic plants are rich in sulfur and volatile oils. It is this combination that makes garlic so valuable as a worm expellent, particularly beneficial to sheep and goats as well as dogs.

Garlic is excellent for the health of cattle but imparts an off-flavor to the milk. There are two ways of reducing this effect. Either feed it at milking time so that the aroma will leave the bloodstream by the time of the next milking, or allow the animal to feed in a pasture with garlic only immediately after milking. Remove her in about an hour and put her on another pasture.

Garlic helps to immunize against infectious diseases. It is helpful in cases of fever, gastric disorders, and rheumatism, and effective against parasites such as ticks, lice, and liver fluke. Garlic is also thought to increase the fertility of animals.

Groundsel (herb of Venus). Groundsel is rich in several minerals, particularly iron. Animals, mainly poultry, seek it out as a tonic. A poultice made of the fresh leaves has antiseptic and drawing properties.

Hop (herb of Mars). This is a tall vine cultivated for its scaly, conelike fruit. Young shoots of hops are much liked by animals and are considered a good conditioner, being tonic and nervine. Hops are also believed to be antiseptic and useful as a vermifuge. The flowers, fed to cows and goats, are a milk stimulant.

Horehound, white (herb of Mercury). Horehound grows easily, and may escape from cultivated gardens to become a weed. The gray-green, woolly leaves are quite bitter. The flowers are small and almost white.

Horehound is one plant on whose medicinal properties practically everyone agrees, and it is probably better known than any other herbal remedy. It is useful as a cough remedy in the treatment of pneumonia, colds, and lung disorders. Horehound used alone may be made into a strong brew. You can also mix it with other herbs such as hyssop, rue, licorice root, and marshmallow root. The addition of honey makes it more palatable. An infusion of horehound is frequently used to make a candy sold in drugstores as "cough drops."

Horsetail (herb of Saturn). Horsetail plant, highly medicinal, is primarily a wound herb, being healing for abscesses and cuts as well as burns. It is soothing and does not sting. Place a handful of the herb, either fresh or dried, in just enough vinegar to cover. Simmer for 20 minutes, cool, and strain. Refrigerate

HORSETAIL BREW

Prepare the brew by placing a handful of the herb, either fresh or dried, in a vessel with just enough vinegar to cover. Allow to simmer for 20 minutes, then cool and strain. Keep it handy in the refrigerator. When needed, add 1 part brew to 2 parts raw milk (use goat's milk if you have it), and pour into a plastic squeeze bottle. Store any leftover brew in the refrigerator.

any portion not immediately used. When needed, add 1 part brew to 2 parts raw milk. If poured into a plastic squeeze bottle, it will be easy to apply.

Hyssop (herb of Jupiter). Grows best in full sun in a fairly dry and medium-rich soil. It is found in waste places. Hyssop is a perennial, and grows about 18 inches tall. It has terminal heads of small, purplish flowers. To use as a tonic, mix 2 teaspoons of herb to a pint of water, simmering about 20 minutes. Hyssop is mildly vermifugal and may be used for delicate lambs and kids.

Dip a poultice of bruised leaves, tied in thin cloth, in hot water and use as an eye treatment.

Knotgrass (herb of Saturn). Knotgrass is a plant of waste places and damp areas. The loosely knotted stems usually recline on the ground at the base. Both its fleshy leaves and pinkish flowers are quite small.

This is primarily a wound plant, applied externally for skin inflammations, abrasions, and ear ailments. Make an infusion of 1 handful of finely chopped herb in a small quantity of water.

Lavender (herb of Mercury). Grow in full sun, in fairly dry and medium-rich soil. The plant is of shrubby habit with gray, fine-leaved foliage with blue, very fragrant flowers.

Lavender, highly tonic, is a great favorite with sheep and goats. It gives a sweet flavor to milk and cheese, is highly antiseptic, and prevents quick souring of milk. The whole plant is useful, especially the flowers. Few ticks are found where lavender forms a ground cover. The crushed leaves are used as an insect repellent and to deter mice.

An infusion of lavender, made by brewing a handful of the herb in water, is good for nervous ailments, vomiting, and sunstroke.

Lemon (herb of the sun). Lemon fruit has unusual healing powers, being especially good as a blood cleanser. Good for fevers, diarrhea, and worms, it may also be used externally for skin ailments, abscesses, ringworm, mange, and to cleanse sores.

Add honey when given internally to make it more palatable. Diluted lemon juice is a good treatment for sore eyes.

Lemon balm (herb of Jupiter). Lemon balm is a common garden plant that sometimes runs wild in the United States. A perennial, it dies down in winter but quickly grows again from the root in spring, reaching about a foot in height. The light green, serrated, and wrinkled foliage when crushed smells and tastes like lemon. It is pleasant and refreshing.

Bees love lemon balm, so plant some in or near apiaries. Rub the inside of a new hive to induce the bees to stay.

Lemon balm is a good pasture plant, promoting the flow of milk in cows and other grazing animals. It's good for retained afterbirth and uterine disorders. Give a tea of lemon balm and marjoram to cows after calving to strengthen and soothe them. Use 2 ounces of each herb (dried or fresh) to 1 pint of water.

Marigold (herb of the sun). Both sheep and goats eat marigold eagerly. It is a good heart medicine and restorative to veins and arteries. Both the leaves and the golden flowers are used. Mix into bran and give twice daily.

Tincture of marigold is good for cuts, bruises, sprains, wounds, warts, and eczema. Crushed blossoms, soaked in vinegar, alleviate pain and swelling caused by bee or wasp stings. Marigolds in the garden act as insect repellents.

Marjoram (herb of Mercury). Grow in full sun, in medium-rich, well-drained soil. The spiky heads of purple-rose flowers are honey-sweet and greatly beloved of bees. The entire plant is highly aromatic and when eaten by sheep and goats imparts a fine flavor to the milk. An excellent tonic, it is nervine and blood purifying, useful in digestive and nervous ailments.

Marjoram is helpful for aches and pains: Cut some of the herb fine, place in thin cloth, dip in hot water, and place over the swollen or inflamed area. Marjoram is good for wounds and abscesses, as well.

Feed marjoram to cows to prevent abortion. A tea of marjoram and lemon balm is useful to cows after calving.

Mint (herb of Venus). There are many varieties of mint: apple mint, curly mint, and orange mint, to name but a few. All like semishade and moist, rich soil and should be cut often. Mints are very aromatic, and beloved of bees. As an aphrodisiac tonic for male animals, mint is particularly good for horses and bulls. It will decrease milk flow and may be used to dry off animals in the last months of pregnancy.

Peppermint alleviates digestive disturbances such as hyperacidity, colic, vomiting, and flatulence. Externally, mint is good for aches and pains. It is also an insect repellent.

Mulberry (herb of Mercury). Both fruits and bark are tonic, that is, blood-cooling and cleansing. Leaves and fruit provide a good treatment for worms, particularly in horses. They are also laxative.

Mulberry

Mustard, black (herb of Mars). Mustard possesses highly disinfectant properties and grown in pastures will cleanse the earth. Animals like to eat mustard and it is tonic for them, promoting appetite and aiding digestion. Musard, made into a poultice, or plaster, has long been a remedy for congestions of the chest. Mustard seeds are considered a vermifuge. White mustard is also used for the same purposes but is less effectual. Mustard contains the minerals iodine and sulfur.

Nettles (herb of Mars). A common plant of waste places and hedgerows, nettles have whitish green flower clusters and stinging, hairy leaves. Neutralize the formic acid on the leaves by rubbing the skin with jewelweed or with any member of the sorrel family, including rhubarb.

Nettles increase blood circulation and act as a stimulant. Nettles are rich in both vitamins and iron and are a remedy against anemia. They strengthen the vitality of all animals. Nettles seem to prevent worms and increase milk yield. Nettle juice will curdle milk to be used for cheese-making.

Animals will not eat growing nettles; they are best cut and dried as nettle hay. Chop and add to other feeds and they will be eaten greedily.

Both cows and goats give better and more nutritious milk when fed on nettle hay. Horses improve in health when fed on nettle hay, and their coats have more shine. The droppings of animals fed on nettle hay are especially rich. Nettles added to the compost stimulate humus formation.

Dandelion greens, mixed with young nettles, help prevent or cure coccidiosis in chicks. Chicks grow out faster if powdered nettle leaves are added to their rations. Mix powdered nettle leaves with mash for hens and their eggs will have a higher nutritional value. Turkeys, like other fowl, will fatten on dried nettles.

Onion (herb of Mars). These bulbous plants belong to the lily family, along with garlic, leeks, shallots, and chives. Onion juice relieves the pain of bee and wasp stings. Onions are highly nutritious, soothing, antiseptic, and vermifugal. They are tonic and cleansing and alleviate symptoms of colds and fevers.

Almost all animals like the many plants of the onion family and benefit from their health-giving properties. However, if milking animals are allowed to graze on onions, whether cultivated or wild, the milk will be flavored.

Parsley (herb of Venus). Grow parsley in full sun to semishade in well-drained, medium-rich soil. This is one of our most important herbs. Sheep and goats love it. It improves their milk yield and is good against foot ills.

Parsley is rich in iron and copper and improves the blood. All parts are useful. The seed is highly tonic. Parsley contains vitamins A and B_1 and is good in cases of rheumatism, arthritis, emaciation, acidosis, and catarrh, and for diseases of the urinary tract. Raw parsley juice is helpful in eliminating poisonous drugs from the body, and parsley tea, a diuretic, acts as a mild sedative.

Pennyroyal (herb of Venus). This herb, which should be cut often, likes semishade and moist, rich soil. Pennyroyal is a good insecticide and mosquito repellent, and may be bruised and rubbed on both humans and animals to keep insects away. Pennyroyal rubbed on your cat's collar will discourage fleas.

Animals seek out pennyroyal, eating it for its tonic and stimulating properties. It is stimulant and restorative fed to cows after calving, and effective against digestive ailments, coughs, and colds.

Raspberry (ruled by Venus). Plant in the second quarter under Cancer, Scorpio, or Pisces. This is a woodland plant that likes to grow near water. It has small, white, roselike flowers, and rose-type leaves with a silvery underside. The brilliant red, black, or yellow fruits are very juicy with small seeds.

Raspberry is a member of the Rose family. The leaves and fruits are well liked by all animals, especially goats. The principle fragrine, contained in the foliage, exerts a strong influence on the pelvic girdle, toning the muscles when administered during parturition. Tonic and cleansing, the foliage also improves the condition of animals during pregnancy, aiding the expulsion of the young at birth. Good for digestive ailments, including diarrhea, raspberry is also useful for young animals.

Raspberry herb is considered tonic for all male animals and is said to be a cure for sterility.

Rose (herb of Jupiter). The flowers, leaves, and fruits are all medicinal and beneficial in the animal diet, important to the maintenance of health, rather than curative. Tonic in its properties and a stimulant to the nervous system, this desirable shrub is both beautiful and useful. It makes an excellent hedgerow.

Rosemary (herb of the sun). Grows best in full sunlight, well-drained, slightly alkaline soil. Rosemary sports beautiful blue flowers and small, dark, shiny, highly aromatic leaves.

Goats and sheep graze rosemary with gusto and it gives a fine flavor to their milk. Rosemary is both tonic and antiseptic. Mixed with salt, it is an effective wound herb. It is also valuable as an insecticide.

Rosemary

Rue (herb of the sun). Many consider rue the most bitter of all herbs. It intensely dislikes basil, one of the sweetest, and the two should not be grown near each other. Rue likes full sun and moist, well-drained, medium-rich soil. It repels insects. Cats dislike rue, so rub it on furniture to keep them from clawing it. In the dog's bed, rue will repel fleas.

Sage (herb of Jupiter). Sage foliage is woolly, gray, and aromatic. The flowers are blue-gray and fragrant, and the nectar makes a delicious honey. Sage is enjoyed by all animals, both wild and domestic, who seek it out for its tonic properties. Sage is beneficial to the milk yield, causing it to be both refreshing and copious.

Externally, sage leaves can be made into a poultice for bruises. Sage is an aid to digestion, making rich dishes more palatable. The oils of sage have a preservative effect on meat. Few ticks are found where sage forms a ground cover.

Gargle with sage and vinegar to relieve pain of ulcerated gums; sage and cloves together relieve toothache.

Shepherd's purse (herb of Saturn). A common plant but one of our most useful, shepherd's purse grows from a rosette and reaches a height of 6 to 18 inches. It gets its name from the small, purse-shaped seedpods.

Small birds like the seeds, which you may collect for them. Poultry like this herb, too, as do all animals. The plant has good astringent properties. A bit of the juice placed on cotton and inserted in the nostrils arrests hemorrhage. Make a strong brew from the herb to bathe wounds.

Sorrel (herb of Venus). Animals like to eat this acid-tasting herb, including the seeds, and it may be fed to cattle as a tonic. Its cooling and soothing qualities make it useful in the treatment of fevers and overheated blood. The crushed leaves are curative for skin ailments. Sorrel neutralizes the sting of nettles.

Southernwood (herb of Mercury). This herb likes full sun and a well-drained location. The small flowers are greenish yellow and the whitish, leathery, gray-green leaves have a sweet, pungent scent. Used externally, southernwood is good against parasites of the skin and hair, such as fleas. It is also useful as a moth repellent and is sometimes placed among lambskins.

Sow thistle (herb of Venus). Animals like this plant, and as it increases both quantity and quality of milk, it is also known as milk thistle. It is rich in minerals, antiseptic, and valuable for reducing fevers. It also eases palpitations of the heart.

Sow thistle is a soil improver. It is also aids in the growth of pumpkins and cucumbers.

Sunflower (herb of the sun). Sunflowers in bloom are visited by bees for both pollen and nectar. The seeds are loved by birds and are often included in feeds for poultry as well. Sunflower seeds are rich in vitamins B_1, A, D, and F.

Sunfower herb produces seeds so tasty that all animals love them.

Thyme (herb of Venus). The creeping variety, lemon thyme, has been naturalized in North America. It is a fragrant, mat-forming plant with small, dark green leaves and heather-purple flowers with a sweet, pungent scent. It is a milk tonic for sheep and goats and they eat it heartily. The oil, thymol, is a worm expellent.

The flowers of thyme are much visited by bees and the plant is often grown near apiaries.

Valerian (herb of Mars). The medicinal properties are very powerful, so use valerian with caution. It is useful in the treatment of worms and acute constipation, as a calmative for nervousnes and hysteria, and as a carminative.

Used externally, oil of valerian is good as a rub for cramps, swollen veins, and arteries.

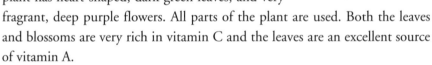

Valerian

Violet (herb of Venus). Small in stature, this plant has heart-shaped, dark green leaves, and very fragrant, deep purple flowers. All parts of the plant are used. Both the leaves and blossoms are very rich in vitamin C and the leaves are an excellent source of vitamin A.

To make a poultice of violet leaves, cover 1 cup of crushed leaves with boiling water. Let stand until cool enough to apply to the affected area. This poultice relieves old wounds, boils, carbuncles, and swellings. Bind loosely and change every few hours.

Watercress (herb of the moon). Watercress grows in shallow water in brooks and ditches. It contains significant quantities of vitamins A, B, C, and B_2, as well as iron, copper, magnesium, and calcium. It promotes strong bones and teeth because of the lime it contains. Good in cases of anemia, it is an excellent blood herb for all animals. Horses and sheep enjoy its foliage and it increases the milk yield of cows and goats.

Willow (herb of the moon). For many years country folk treated fevers with a decoction of willow bark but it was not until the 1820s that its active principle, salicin, was isolated. (In 1899 a synthetic derivative gave the world aspirin.)

Willow is a refrigerant herb, tonic, astringent, antiseptic, and still useful in fevers. Cattle and horses enjoy the young shoots and foliage but it is the bark that contains the crystalline salicin. Flaked bark, brewed in water, is good internally for intestinal inflammations and externally may be used as a massage or poultice.

Witch hazel (herb of Mars). Witch hazel contains tannin, gallic acid, bitter principle, and volatile oils. These are strongly astringent and helpful to stop bleeding both internally and externally.

Yarrow (herb of Venus). Sheep like yarrow and will seek it out in dry areas. Yarrow hay and yarrow tea are also beneficial to sheep. Yarrow is a wound herb good for stopping excess bleeding. Rub on crushed foliage of yarrow to drive away mosquitoes; this also alleviates the sting from bites.

HERBAL PEST CONTROL

To repel *flies,* a decoction of black walnut leaves soaked overnight, an infusion of pignut leaves, or freshly cut pumpkin or squash leaves rubbed on horses or cattle work well. Yellow wild indigo placed in the harness keeps horses free of flies. Rue repels stable flies, so grow it in a border around the barns and poultry house.

Cattle like to rest under nut trees of all kinds, being less troubled with flies there. They also enjoy nibbling on hazelnut trees or bushes in their pasture. The leaves increase the butterfat content of their milk and the tannic acid in the leaves will effectively cleanse their digestive systems. Grow nut trees of all kinds near barns and around manure piles to repel stable flies.

To repel *lice,* rub water in which potatoes have been cooked and pulverized on cattle. Clove will also do this for dogs and chickens.

A Brazilian remedy for poultry lice is a tincture of cocoa leaves; cocoa shells used as bedding for dogs will repel fleas. Try black walnut leaves in the bed of a dog or cat for the same purpose. To rid them of *fleas,* bathe small animals in a water solution of wormwood.

Mosquitoes are repelled by the odor of sassafras; freshly crushed pennyroyal leaves may be rubbed into hair or skin for the same purpose.

Tansy repels *flies and ants* and may be used around both the stable and the house.

There are several ways to relieve cattle from the misery of *ticks.* Few ticks are found where sage or pyrethrum forms a ground cover, and this is also true of lavender. The real prize, though, is molasses grass. Where this is grown, cattle are almost entirely free of these pests. The leaves of this grass are covered with glandular hairs that exude a viscous oil, giving the grass marked adhesive properties capable of trapping small insects. The sticky secretion does not kill young ticks but simply prevents them from crawling upward and coming in contact with an animal.

Besides being a tick deterrent, molasses grass repels ***mosquitoes and tsetse flies.*** And cattle like to eat the grass; in the warmer areas of this country, it has great possibilities.

Grain stored as feed for animals is ofen pilfered by ***mice and rats,*** but dwarf elder will drive them away. So will any of the spurges.

It is possible for you to grow many of the herbs you have read about here. Consult seedsmen and nurseries specializing in herb seeds and plants. For barks and roots, not so easily obtainable, Indiana Botanic Gardens is an invaluable resource.

APPENDIX

SUGGESTED READINGS

Culpeper, Nicholas, *Culpeper's Herbal Remedies*. Beverly Hills, CA: Wilshire Publishing, 1980.

Daath, Henrich, *Medical Astrology. Medical Astrology*. Kila, MT: Kessinger Publishing Company, 1996.

Duz, M. D., *Practical Treatise of Astral Medicine and Therapeutics*. Mokelumne Hill, CA: Helath Research, 1996.

Gettings, Fred, *The Book of the Zodiac: An Historical Anthology of Astrology*. London, Engl.: Triune Books, 1973.

Gibson, Walter B., Gibson, Litzka R., Keshner, Murray, *The Complete Illustrated Book of the Psychic Sciences*. New York: Pocket Books, 1969.

Klober, Kelly, *A Guide to Raising Pigs*. Pownal, VT: Storey Books, 1997.

Moon Sign Book and Gardening Almanac. St. Paul, MN: Llewellyn Publications. (Published annually since 1906.)

Riotte, Louise, *Catfish Ponds and Lily Pads*. Pownal, VT: Storey Books, 1997.

Root, A. I., and E. R., *ABC & XYZ of Bee Culture*. Medina, OH: A. I. Root Company, 1997.

Spence, Lewis, *An Encyclopedia of Occultism*. New York: Citadel Press, 1993.

Stowe, Lyman E., *Stowe's Bible of Astrology*. Kila, MT: Kessinger Publishing Company, 1997.

RESOURCES

American Rabbit Breeders
  Association, Inc.
8 Westport Court
Bloomington, IL 61704
309-664-7500
members.aol.com/arbanet/web

Hoegger Suppy Company
P.O. Box 331
160 Providence Road
Fayetteville, GA 30214
800-221-4628
hoegger@mindspring.com
Mail-order dairy goat supplies.

Indiana Botanic Gardens
3401 W. 37th Avenue
Hobart, IN 46342
800-644-8327
www.botanichealth.com/
*Mail-order herbs and medicinal
products.*

Llewellyn Publications
84 S. Wabasha Street
St. Paul, MN 55107
800-843-6666
www.llewellyn.com
Publisher of The Moon Sign Book
and other astrology titles.

McMurray Hatchery
191 Closz Drive
Webster City, IA 50595
515-832-3280
*Chicks, ducks, geese, guineas,
pheasants.*

Northwoods Publishing
  Company, Inc.
U.S. Highways 41–45
P.O. Box 609
Menomonee Falls, WI 53051
*Offer solunar tables. Search the
Internet for additional sources.*

A. I. Root Company
623 West Liberty Street
Medina, OH 44256
330-725-6677
*Apiary and beekeeping information
and supplies.*

Spring Hill Nurseries
110 W. Elm Street
Tipp City, OH 45371-0169
937-667-4079
www.springhillnursery.com
Mail-order nursery.

INDEX

Note: Numbers in *italic* indicate illustrations; numbers in **boldface** indicate charts.

OTHER STOREY TITLES
YOU WILL ENJOY

Astrological Gardening: The Ancient Wisdom of Successful Planting and Harvesting by the Stars by Louise Riotte. Astrological Gardening has been practiced for centuries, and author Louise Riotte's 40 years of planetary planting have confirmed its usefulness. She identifies ideal astrological conditions for planting and harvesting a wide range of fruits, vegetables, flowers, and herbs. 224 pages. ISBN 0-88266-561-8.

Carrots Love Tomatoes: Secrets of Companion Planting for Successful Gardening (2nd ed.) by Louise Riotte. This classic gardening book is now available in a completely revised edition to inspire and instruct a new generation of gardeners. One of our all-time best-selling books, it lists hundreds of plants and their ideal (and not so ideal) companions. 224 pages. ISBN 1-58017-027-7.

Catfish Ponds & Lily Pads: Creating and Enjoying a Family Pond by Louise Riotte. Riotte explains siting a pond, maintaining water quality, troubleshooting, stocking with fish, plus plenty of old-time fishing lore and scrumptious fresh fish recipes. 160 pages. ISBN 0-88266-949-4.

Roses Love Garlic: Companion Planting and Other Secrets of Flowers (2nd ed.) by Louise Riotte. This sequel to *Carrots Love Tomatoes* lists hundreds of herbs and flowers, with information on how their proximity can maximize the health and yield of vegetables, berry bushes, and fruit and nut trees. The new edition features a dozen illustrated garden plans. 256 pages. ISBN 1-58017-028-5.

Sleeping with a Sunflower: A Treasury of Old-Time Gardening Lore by Louise Riotte. Arranged month-by-month, this book is packed with pure gardening folk wisdom: planting by the moon, fishing when and where the bass will bite, and more. A treasured gift for anyone interested in the forgotten arts, crafts, and skills of early American life. 224 pages. ISBN 0-88266-502-2.

These books and other Storey Books are available at your bookstore, farm store, garden center, or directly from Storey Books, Schoolhouse Road, Pownal, Vermont 05261, or by calling 1-800-441-5700. Or visit our website at www.storeybooks.com.